T0211466

THE BEDFORD SERIES IN HISTORY AND CULTURE

Mao Zedong
and China's Revolutions
A Brief History with Documents

Related Titles in
THE BEDFORD SERIES IN HISTORY AND CULTURE
Advisory Editors: Natalie Zemon Davis, *Princeton University*
Ernest R. May, *Harvard University*
Lynn Hunt, *University of California at Los Angeles*
David W. Blight, *Amherst College*

The Japanese Discovery of America: A Brief History with Documents
Peter Duus, *Stanford University*

Schools and Students in Industrial Society: Japan and the West, 1870–1940
Peter N. Stearns, *Carnegie Mellon University*

Pearl Harbor and the Coming of the Pacific War: A Brief History with Documents and Essays
Akira Iriye, *Harvard University*

My Lai: A Brief History with Documents
James S. Olson, *Sam Houston State University,* and Randy Roberts, *Purdue University*

Mao Zedong and China's Revolutions

A Brief History with Documents

Timothy Cheek

University of British Columbia

palgrave

MAO ZEDONG AND CHINA'S REVOLUTIONS, by Timothy Cheek

The Library of Congress has catalogued the paperback edition as follows:
2001097845

Copyright © Bedford/St. Martin's 2002

Softcover reprint of the hardcover 1st edition 2002 978-0-312-29429-8

PALGRAVE, 175 Fifth Avenue, New York, NY 10010

First published by PALGRAVE, 175 Fifth Avenue, New York, NY 10010. Companies and representatives throughout the world. PALGRAVE is the new global imprint of St. Martin's Press LLC Scholarly and Reference Division and Palgrave Publishers Ltd. (formerly Macmillan Ltd.).

Manufactured in the United States of America.

7 6 5 4 3 2

f e d c b a

ISBN 978-1-349-63485-9   ISBN 978-1-137-08687-7 (eBook)
DOI 10.1007/978-1-137-08687-7
Cover art: *Mao Zedong* © Bettmann/CORBIS.

Acknowledgments

Acknowledgments and copyrights can be found at the back of the book on pages 243–44, which constitute an extension of the copyright page.

Transferred to Digital Printing 2008

Foreword

The Bedford Series in History and Culture is designed so that readers can study the past as historians do. The historian's first task is finding the evidence. Documents, letters, memoirs, interviews, pictures, movies, novels, or poems can provide facts and clues. Then the historian questions and compares the sources. There is more to do than in a courtroom, for hearsay evidence is welcome, and the historian is usually looking for answers beyond act and motive. Different views of an event may be as important as a single verdict. How a story is told may yield as much information as what it says.

Along the way the historian seeks help from other historians and perhaps from specialists in other disciplines. Finally, it is time to write, to decide on an interpretation and how to arrange the evidence for readers.

Each book in this series contains an important historical document or group of documents, each document a witness from the past and open to interpretation in different ways. The documents are combined with some element of historical narrative—an introduction or a biographical essay, for example—that provides students with an analysis of the primary source material and important background information about the world in which it was produced.

Each book in the series focuses on a specific topic within a specific historical period. Each provides a basis for lively thought and discussion about several aspects of the topic and the historian's role. Each is short enough (and inexpensive enough) to be a reasonable one-week assignment in a college course. Whether as classroom or personal reading, each book in the series provides firsthand experience of the challenge— and fun—of discovering, recreating, and interpreting the past.

Natalie Zemon Davis
Ernest R. May
Lynn Hunt
David W. Blight

Preface

What can the life of a notable figure tell us about the experience of a people or a country? And how can we get at that life—get access to the thoughts and goals of a leader? This book seeks to provide such an avenue through a brief history and documents by and about Mao Zedong. Mao is probably the greatest figure of twentieth-century China—a hero to some, a demon to others. He led the Chinese Communist party (CCP) to national victory in 1949. He drove the People's Republic of China (PRC) through three decades of tumultuous revolutions, from the Soviet model in the 1950s, to the Cultural Revolution in the 1960s, to rapprochement with the United States in the 1970s. More than twenty-five years after his death in 1976, Mao remains an iconic figure in China today, both as the indispensable legitimator of the troubled CCP and as the object of popular fascination and nationalist hopes.

Mao's writings provide a concrete path into the experience of China's twentieth-century revolutions, as he deals with rural misery in the 1920s, government formation in the 1940s, and social revolution in the late 1950s and 1960s. Earlier histories, in the West as well as in China, have conventionally presented Mao as the embodiment of each stage of *the* Chinese revolution. Now we can ask, When was Mao in synch or out of synch with the social experiences and political aspirations of major groups in China? In the 1920s, Mao was not the most important CCP leader, but his 1927 "Report on the Peasant Movement in Hunan" was an accurate assessment of rural poverty and its potential as a catalyst for social revolution. By 1940, when Mao wrote "On New Democracy," he was a top leader of a revived CCP, and the plan outlined in that essay became the public blueprint for the CCP's takeover of China in 1949. By 1957 Mao was supreme leader and ideological fountainhead, but his assessment of Chinese society, as well as his hypocritical reluctance to follow his own prescriptions, made his "On the Correct Handling of Contradictions among the People" both

vi

misrepresent reality and contribute to tragedy. By the era of the Cultural Revolution, beginning in 1966, Mao's writings had been reduced to oracular pronouncements and sound bites from the famous "Little Red Book" *(Quotations from Chairman Mao Zedong)*. These writings did not provide an accurate assessment of Chinese society, but they did contribute to the Mao cult and the emerging chaos of the Cultural Revolution.

The three long essays by Mao referred to above characterize three periods in the life of Mao and China. Briefer selections of the Chairman's writings are also included to provide a sense of Mao's nationalism, poetry, literary policies, and leadership model. In each of these three periods—rural revolution in the 1920s and 1930s, political revolution in the 1940s, and utopian social revolution in the late 1950s and 1960s—we can observe Mao's contributions (for good or for ill) to China's varied revolutions in nationalism, socialism, and economic development.

This collection offers students in an introductory or world studies course a manageable and representative sample of Mao's writings. I have chosen long extracts from the three periods outlined above. As a result, I have omitted some of the topics covered by Mao's extensive corpus, such as his views on the Soviet Union and land reform, his philosophical essays, and his economic plans. These are all available in the new comprehensive collections of Mao's writings edited by Schram and by Kau and Leung (see the bibliography). Some teachers may find the three core texts too long and may choose to focus on certain sections of them. Or they may focus on one of the three main essays (and the related secondary texts in part two) or on certain themes, such as the status of women or changes in rural life. In any case, these three core texts represent complete thoughts by Mao that were published under his scrutiny and were influential both inside and outside China. They remain important primary documents, which intelligent readers can mine for purposes beyond those I suggest.

The Mao texts are followed by several writings about Mao to help describe the contexts in which Mao operated and to indicate something of what Mao meant to Chinese and non-Chinese in the twentieth century. They range from Edgar Snow's famous interview with Mao in 1936, to the memoirs of his doctor, to uses of Mao by Red Guards in the Cultural Revolution and by people in China today. This section also includes a taste of what academics have tried to contribute to our understanding of Mao. The romanization of Chinese names and words has been changed to the pinyin system, except for a few names, such as Chiang Kai-shek and Sun Yat-sen.

The volume begins with a comprehensive introduction, which presents the major issues in modern Chinese history and Mao's growing role in the events of the twentieth century. It includes a small set of photographs and graphic images to give a sense of how Mao was portrayed. In all, this book aims to equip students to make their own readings of Mao's writings and to find for themselves what the "Great Helmsman's" life and work can teach us about China's continuing revolutions.

ACKNOWLEDGMENTS

I wish to thank Jeff Wasserstrom for making this project possible and Louise Townsend for making it happen. I am grateful to the students in my Mao seminar at Colorado College for testing the readings and offering necessary suggestions. I owe a special debt to the five reviewers for Bedford/St. Martin's—Stephen Averill, Jeff Hornibrook, Steven I. Levine, Sandra Loman, and Patricia Stranahan—for their alarmingly detailed and trenchant readings of the first draft of this book. Although I was not able to follow all of their suggestions, to the degree that this book achieves the impossible goal of satisfying both specialist and newcomer, it is due to our collective efforts. Thanks as well to Tim Brook, Nancy Hearst, Nick Knight, David Ownby, Stuart Schram, and Mark Selden for making careful readings of earlier drafts and providing extensive comments and suggestions. Also thanks to the following colleagues for helpful suggestions: Vera Fennell, Anne Hyde, David Kelly, Tony Saich, and Michael Schoenhals. I am grateful to Stuart Schram not only for his prodigious scholarship, on which I have relied, but also for permission to use the forthcoming translation of "On New Democracy" from his *Mao's Road to Power* collection. Sandy Papuga helped bring the manuscript to order, and Jack Hayes researched the photos used in this volume. The professional team at Bedford/St. Martin's made this the best book it can be under the guiding hand of Emily Berleth. Nancy Benjamin at Books By Design served as efficient project manager, Rachel Siegel assisted with permissions, and Billy Boardman created the striking cover art. Special thanks are due to Barbara Jatkola for copyediting above and beyond the call of duty. Finally, thanks to my family for their patience, and especially to my daughter, Tessa.

                                                     Timothy Cheek

Contents

Maps and Illustrations

Introduction:
Comrade, Chairman, Helmsman —
The Continuous Revolutions of Mao Zedong

When Mao Zedong was thirteen, he had a huge fight with his strict father during a large party at their home. Mao recalled later, "My father denounced me before the whole group, calling me lazy and useless. This infuriated me. I cursed him and left the house. My mother ran after me and tried to persuade me to return. My father also pursued me, cursing at the same time that he commanded me to come back." But Mao was sick of his father's harsh treatment and refused to obey. In fact, Mao threatened to jump in the nearby pond and kill himself. Faced with this fierce resistance, Mao's father gave in. They agreed: Mao would obey his father if his father would promise to stop beating him. "Thus the war ended," Mao said, "and from it I learned that when I defended my rights by open rebellion my father relented, but when I remained meek and submissive he only cursed and beat me the more."[1] Mao carried this lesson into political life. He always sought revolutionary answers to the social and political problems he encountered. He was the continuous revolutionary—from his early days as a radical student in rural China in the 1910s to his last days as supreme leader in Beijing in the 1970s.

Mao Zedong lived from 1893 to 1976. He is the most famous Chinese of the twentieth century and certainly China's most influential political leader. Mao has been presented by his faithful followers in the Chinese Communist party (CCP) as the embodiment of China's socialist revolution, which brought the Communists to power in the new People's Republic of China (PRC) in 1949. That "story" was extremely compelling—not only to a majority of Chinese over several decades at mid-century, but also to non-Chinese academics trying to explain

[1]As told to Edgar Snow in *Red Star over China* (New York: Grove Press, 1968), 133. See extracts of Snow's interviews with Mao in 1936 in Document 11.

1

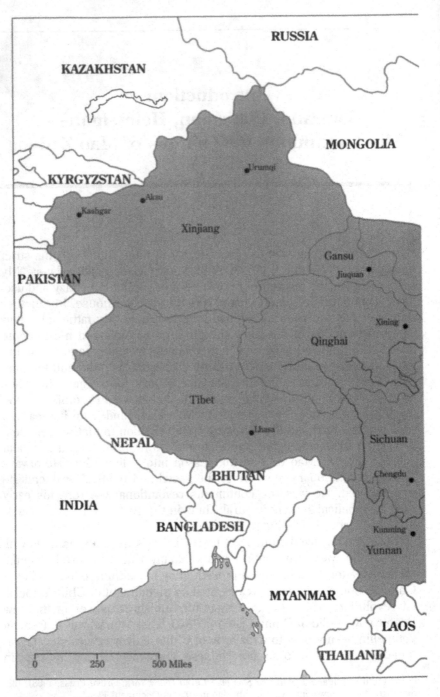

RUSSIA

KAZAKHSTAN

MONGOLIA

KYRGYZSTAN

Urumqi

Aksu

Kashgar

Xinjiang

Gansu

Jiuquan

PAKISTAN

Xining

Qinghai

Tibet

Lhasa

Sichuan

NEPAL

Chengdu

BHUTAN

INDIA

BANGLADESH

Kunming

Yunnan

MYANMAR

LAOS

0   250   500 Miles

THAILAND

Map 1. *China Today*

2

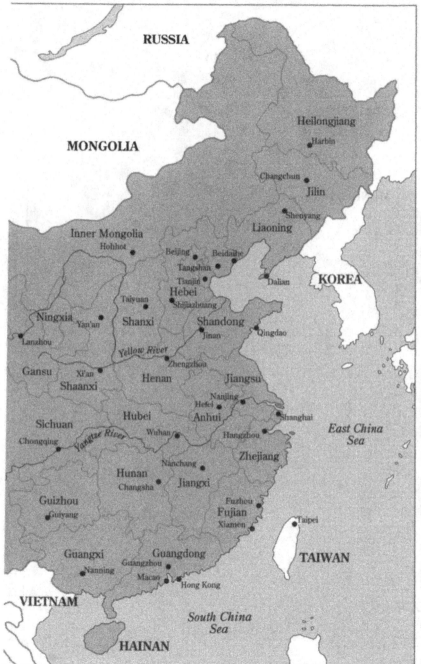

RUSSIA

MONGOLIA

Heilongjiang
Harbin

Changchun
Jilin

Shenyang

Inner Mongolia
Hohhot

Liaoning

Beijing   Beidaihe

Tangshan

Tianjin

Taiyuan

Hebei

Dalian

KOREA

Shijiazhuang

Ningxia
Yan'an

Shanxi

Shandong

Lanzhou

Yellow River

Jinan

Qingdao

Gansu
Xi'an

Zhengzhou

Shaanxi

Henan

Jiangsu

Nanjing

Hefei

Sichuan

Hubei

Anhui

Shanghai

Chongqing

Yangtze River

Wuhan

Hangzhou

East China
Sea

Zhejiang

Nanchang

Hunan
Changsha

Jiangxi

Guizhou
Guiyang

Fuzhou

Fujian

Xiamen

Taipei

Guangxi
Nanning

Guangdong

Guangzhou

TAIWAN

Macao

Hong Kong

VIETNAM

South China
Sea

HAINAN

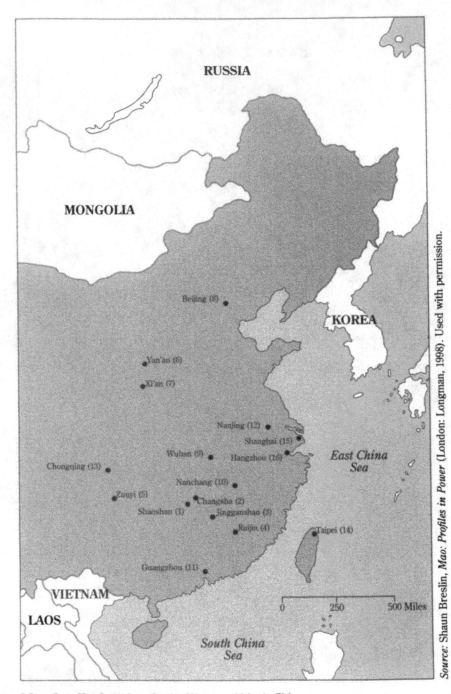

RUSSIA

MONGOLIA

Beijing (8)

KOREA

Yan'an (6)

Xi'an (7)

Nanjing (12)

Shanghai (15)

Wuhan (9)　Hangzhou (16)

Chongqing (13)

East China
Sea

Nanchang (10)

Zunyi (5)

Changsha (2)

Shaoshan (1)　Jingganshan (3)

Ruijin (4)　Taipei (14)

Guangzhou (11)

VIETNAM

LAOS

0　　250　　500 Miles

South China
Sea

Source: Shaun Breslin, *Mao: Profiles in Power* (London: Longman, 1998). Used with permission.

Map 2. *Key Locations in the History of Mao's China*

Notes to Map 2

(1) Shaoshan: Mao's birthplace.

(2) Changsha: Where Mao went to school and later worked in the Changsha Normal School. Also the site of Mao's abortive "Autumn Harvest" uprising.

(3) Jingganshan: The original safe haven in Jiangxi province that Mao and others fled to after the failed uprising of 1927.

(4) Ruijin: The capital of the Jiangxi soviet.

(5) Zunyi: The site of the conference in 1935 where Mao began his rise to supreme power in the CCP.

(6) Yan'an: The Communist headquarters after the Long March and throughout the war against Japan.

(7) Xi'an: The site of Chiang Kai-shek's kidnapping by the northern warlords, which led to the creation of the second united front.

(8) Beijing: China's capital before the collapse of the empire in 1911 and once again after the revolution.

(9) Wuhan: Birthplace of the 1911 Revolution and the site of intramilitary conflict at the height of the Cultural Revolution. Also briefly twice China's capital under the Guomindang—first as they moved north before taking Nanjing, and later as the Guomindang fled Nanjing after the Japanese invasion.

(10) Nanchang: Site of a doomed Communist uprising in 1927.

(11) Guangzhou (Canton): The Guomindang's headquarters after Yuan Shikai ignored the national election result in 1912. The scene of many and bitter battles during the 1920s. Also the site of a short-lived commune after the Communist uprising in December 1927.

(12) Nanjing: Capital of China after the 1911 Revolution, and again after the Guomindang established a new national government after the Northern Expedition. Site of the "Rape of Nanjing" after the Japanese invasion and of the puppet regime of Wang Jingwei under Japanese control.

(13) Chongqing: Became the Guomindang's base during the war against Japan.

(14) Taipei: The capital of the Guomindang's Republic of China from 1949 to the present day.

(15) Shanghai: China's major commercial center and one of the centers of radicalism during the Cultural Revolution. The basis of the Gang of Four's power in the early 1960s.

(16) Hangzhou: Another center of radical leftism. Site of a number of important conferences and meetings at the outset of the Cultural Revolution.

China's many social and political changes.[2] It is not so simple now. Mao's faults, the achievements of other CCP leaders, and the contributions of talented Chinese in other arenas are all part of the story of China's several revolutions. The question, then, is how best to use Mao to understand China's revolutions.

[2]Mark Selden, *The Yenan Way in Revolutionary China* (Cambridge: Harvard University Press, 1971). Even those critical of Mao put him at the center of a single Chinese revolution. See Merle Goldman, *Literary Dissent in Communist China* (Cambridge: Harvard University Press, 1967), and Simon Leys, *Chinese Shadows* (New York: Viking, 1977).

This reader seeks to provide students with the means to use writings by and about Mao Zedong to understand China's twentieth-century revolutions. Three of the most important revolutions in which China was engaged were nationalism, socialism, and economic development. Each took root during the twentieth century, and as leader of the CCP, Mao was active in each. One way to begin to use Mao's writings and work as a window on China is to ask, When was Mao in synch or out of synch with the social experiences and political aspirations of major groups in China? This requires us to have some sense of what Mao was thinking and what those around him, as well as people throughout China, were experiencing. The three long texts by Mao and excerpts from some of his other essays in part one, along with the related readings by various Chinese and Western writers in part two, offer sufficient information for the careful reader to approach this goal. They provide some understanding of China's revolutions in nationalism, socialism, and economic development; of Mao's contributions to those revolutions; and of what it meant to be a Chinese revolutionary—one of the millions who followed Mao from dispirited poverty to unimaginable successes and then to unanticipated disasters. This mixed record continues to shape Chinese culture and society today.

CHINA'S REVOLUTIONS

China's leaders today wish to fit in with the international political and economic system, but they wish to do so with "Chinese characteristics." Whereas the leaders of the CCP seek "socialism with Chinese characteristics," the rival Chinese government on Taiwan, run until recently by the Guomindang (GMD), or Nationalist party, has worked to preserve Chinese identity while engaging in global capitalism. Both efforts were part of China's twentieth-century revolutions, which transformed it from a Confucian empire to a modern nation-state.

Mao came to represent the efforts of many Chinese to address the challenges of nationalism, socialism, and economic development, and he brought China a long way toward achieving the success it enjoys today. But in the 1920s, few people thought that China could solve its problems of internal disorder and foreign encroachment, and nearly no one had heard of Mao Zedong.

When Mao was born in 1893, China's Qing dynasty was faltering. In 1895 the empire lost a major war with Japan. Five years later, it fell victim to the Boxer Rebellion of 1900 and the occupation of the capital, Beijing, by European and Japanese forces, which put down the revolt. Finally, the empire collapsed in the 1911 Revolution, which promised a constitutional republic but delivered only militarist rule and competing warlords across China. During these years, Mao grew up as the eldest son of a prosperous farmer in the central Chinese province of Hunan. His father was a harsh patriarch and his mother a devout Buddhist. Mao studied the ancient Confucian classics at a local school but took a greater interest in the new writings of Chinese reformers and anarchists. Meanwhile, his father wanted him to learn the family business. This led to a violent confrontation between father and son, and Mao took his rebellious attitude into the cultural turmoil and political debate of the 1910s.

China in the 1910s and 1920s was in trouble. Politically, economically, and socially, it was on the verge of utter collapse—or at least it appeared to be so. Politically, China had failed in its attempt to turn a dynastic empire into a nation-state run as a constitutional republic. Economically, foreign commercial interests plundered China's raw materials and flooded it with manufactured goods that drove rural handicraft households into bankruptcy. Socially, the links between scholars and the state and between scholar-gentry families and farming communities had been broken, leaving the government to fools and villains and the rural order to thugs.[3] Mao's own account of his youth (see Document 11) gives a vivid sense of these troubled years. Yet these same years were a time of vibrant intellectual and political experimentation, and Mao contributed to that vitality.

From around 1915 on, Chinese scholars and writers increasingly saw the previous century as one long story of decline and failure in China. They worried that China might cease to exist. Since the 1911 Revolution had not made China strong and independent, these scholars came to believe that China needed a "new culture" to restore its glory.

They had a point. In the eighteenth century, China had been a proud and militarily expansionist empire with a vibrant commercial economy and rich urban culture.[4] Yet it was also overpopulated, served by an understaffed bureaucracy, and unwilling to deal with

[3]Two excellent textbooks are Jonathan Spence, *The Search for Modern China*, 2nd ed. (New York: W. W. Norton, 2000), and John King Fairbank and Merle Goldman, *China: A New History* (Cambridge: Harvard University Press, 1998).

[4]Susan Naquin and Evelyn Rawski, *Chinese Society in the Eighteenth Century* (New Haven: Yale University Press, 1987).

upstart strangers at its back door—British merchants on the southeast coast. China had maintained commercial links with "Nanyang," the southern seas of Southeast Asia, for well over a thousand years. Threats to its national security, however, had always come from nomads in the north and west,[5] including the Mongols, who successfully occupied all of China in the thirteenth century, and the Manchus of the Qing dynasty, who ruled from 1644 to 1911. Those invaders had never questioned the cultural-political model of imperial Confucianism. They had added their own practices to Chinese culture, but they largely adopted Chinese institutions and tended to become Chinese in their habits and interests. This was not the case with these new barbarians from the sea.

From the beginning of direct contact with Europeans in the sixteenth century, China could tolerate Catholic missionaries and trading sea captains from a position of strength. But the decline of Qing administrative efficiency and military might tragically coincided with the unprecedented growth of European means and motives to get wealth out of China and to "save" China's residents from their ignorance of Christianity and Europe's supposedly superior civilization. Never had China been so profoundly challenged on all three levels of strategic order, economic organization, and social and cultural values.

The Qing dynasty buckled under the weight. The Opium War of 1839–42 was the turning point. After that defeat to British naval power and commercial zeal, the Qing conceded five "treaty ports" where first British and then all foreign merchants and diplomats could set up shop—under their own law. From the foothold of these treaty ports, Western ideas and practices began their slow and unsettling challenge to Chinese government and ways of life. The Qing never really recovered. The devastating Taiping Rebellion of 1850–64, fomented by a bizarre pseudo-Christian sect,[6] engulfed more than half of China. Suppressing the Taiping rebels required the Qing to give more treaty ports to the Western powers and, fatefully, more military independence to regional Chinese loyalists.

The Qing finally collapsed in the 1911 Revolution, to be replaced by a short-lived republic under Sun Yat-sen, China's revered father of modern revolution. Sun was a republican revolutionary and is considered the founder of the Nationalist party (the GMD). However, Gen-

[5]Sechin Jagchid and Van Jay Symons, *Peace, War, and Trade along the Great Wall: Nomadic-Chinese Interaction through Two Millennia* (Bloomington: Indiana University Press, 1989).
[6]Jonathan Spence, *God's Chinese Son* (New York: W. W. Norton, 1996).

eral Yuan Shikai, China's first warlord, quickly forced him to resign. China sank into what Mao would later call semifeudal, semicolonial conditions, where no real central government held sway over more than a few of China's twenty-nine provinces and the rest were controlled by a motley crew of militarists and foreign powers.

BATTLE CRY

It was these dire straits that provoked a sense of terminal crisis among China's scholars and writers and led to the New Culture Movement (1915–25). Under the empires, China's scholars were versed in the Confucian canon, vetted by state examinations, and valued as leaders of government and society. That link, however, was broken by 1905, when the Qing discontinued the state examination system, and hopes for its revival or replacement foundered with the fall of the Qing dynasty and the sinking of the republic. On top of all this, China's patriots had to watch Japan—a traditional tributary to China's courts—swagger in as the newest foreign imperialists, taking what they pleased from China. When the Versailles treaty of 1919 gave former German territories in China to the Japanese instead of returning them to China (which, like Japan, had fought with the allies), China's intellectuals hit the streets in protest.

The archetype of militant protests in modern China was the May Fourth Movement.[7] On May 4, 1919, Beijing University students and many others protested the Versailles treaty and forced the Chinese delegation not to sign it (China never did). This political movement gave focus to the broader New Culture Movement. Between 1915 and the mid-1920s, China's new urban educated class tried to reform China's high culture so as to save it from extinction. The essays and magazines Mao read as a youth came from the New Culture Movement, especially Liang Qichao, whose writings were first published at the turn of the century. Chen Duxiu's radical *New Youth* journal was required reading. Hu Shi and other scholars led the vernacularization effort—the shift from the terse and complex classical Chinese of *wenyan* to the easier-to-read *baihua*, which mimicked the spoken language and helped more people learn to read and write.

The battle cry of the New Culture Movement, as Mao recalls, was "Down with the house of Confucius." Yet the West, with its model of

[7]See Jeffery Wasserstrom, *Student Protests in Twentieth-Century China* (Stanford: Stanford University Press, 1991).

parliamentary democracy, was by 1920 neither the only nor the most
compelling alternative to stuffy Confucian rituals. Sullied by their association with economic and political imperialism, Western liberal ideals
were utterly discredited by the devastation of World War I. Thousands
of Chinese students who worked as trench diggers and factory hands
in France got a firsthand look at that European disaster. By 1917
China was ripe for the news of the October Revolution in Russia, in
which the Communist Bolsheviks led by Lenin toppled the old tsarist
regime. By 1919 Marxism, which had been known but not popular in
China for a decade, became a viable alternative to what became known
in Mao's view as feudalism (Chinese tradition) and imperialism (Western liberal democratic models).

The CCP began to take shape in 1920 in at least four spots around
China, ranging from Guangzhou to Sichuan.[8] Mao was an early partic-
ipant in one of the regional study groups—the New People's Society
in Changsha, Hunan province, where he owned a radical bookstore.
He served as one of two Hunan delegates to the First Party Congress
of the CCP in Shanghai in July–August 1921.[9] From the start, the
party was guided by agents of the Comintern—Lenin's Communist
International organization. These Soviet agents oversaw the fledgling
Chinese leftists in the creation and operation of a Bolshevik party.
This brought the prodigious power of Leninist "party cell" organization and the mobilizing ideology of Marxism-Leninism into China.[10]
The fit was not good. Soviet theory focused on the industrial working
class, or proletariat, but the proletariat in China was tiny. Furthermore, Soviet state policy conflicted with its ideological program. At the
same time it helped the CCP, the USSR in fact simultaneously supported the rival Nationalist party, or GMD, as a hedge to guarantee a
Chinese government that would be friendly to the Soviets.

The CCP struggled with the GMD, although at first the two worked
together. Sun Yat-sen, the erstwhile first president of the republic, had

---

[8]Hans J. van de Ven, *From Friend to Comrade: The Chinese Communist Party,
1920–1927* (Berkeley: University of California Press, 1991), chap. 1.

[9]The best documentary collection and analysis of CCP history is Tony Saich, *The
Rise to Power of the Chinese Communist Party: Documents and Analysis* (Armonk, N.Y.:
M. E. Sharpe, 1996). On Mao's skill as a bookstore owner, see Jonathan Spence, *Mao
Zedong* (New York: Viking Penguin, 1999), 44.

[10]On the considerable organizational powers of Leninism and its contributions to
national independence struggles and economic modernization efforts, see Kenneth
Jowitt, "The Leninist Response to National Dependency," in Kenneth Jowitt, *New World
Disorder: The Leninist Extinction* (Berkeley: University of California Press, 1992), 1–50.
On the Comintern and the Chinese revolution, see Tony Saich, *The Origins of the First
United Front in China* (Leiden: E. J. Brill, 1991).

retreated after his forced resignation in 1912 to form the GMD, a revolutionary party. By the mid-1920s, it had a foothold in southern China but needed support. Only the new Soviet Union would assist Sun against the northern warlords. The price was a "united front"—political cooperation with the fledgling CCP. When Sun died in 1925, Chiang Kai-shek succeeded him as leader of the GMD.

During the Northern Expedition of 1926–27, forces under Chiang worked with the Communists to reclaim nearly half of China's territory—from Guangdong in the south to central China and even Beijing (see Map 1)—under a central government. Mao refers to the soldiers in this expeditionary force in his 1927 "Report on the Peasant Movement in Hunan" (see Document 1). But the alliance between the CCP and GMD soon crumbled. In April 1927, Chiang, who had visited Soviet Russia and understood the expansionist nature of the Communist party, turned on the CCP in a surprise attack that wiped out the urban party infrastructure.

Into this void stepped Mao Zedong. As outlined in his 1927 report, Mao looked not to the urban revolution of factory workers—the proletariat of Marx and Lenin—but rather to the rural poor, whom Marx had dismissed as "peasants" (see Figure 1). This essay reflects the new conceptual tools that Russian Marxism-Leninism provided Chinese leftists to make sense of their own situation. Mao's breakdown of the rural classes into poor, middle, and rich peasants has become the authoritative statement of Marxist class analysis of the Chinese countryside. Likewise, his description of the peasant associations reflects an attention to the sinews of local power that is a hallmark of Leninism. Mao's genius was to locate the fundamental battleground of this class struggle not at the national or provincial level, but in local society. This view was not, however, popular with either most of his Communist comrades, who believed the proletariat to be the key group, or his Nationalist allies, who did not favor rural insurrection.

Mao's report also reflects the astonishing feminism of some early CCP policy.[11] Mao was certainly a leader in advocating women's liberation from the "masculine authority of husbands," as well as clan, temple, and general religious oppression. That Mao and the CCP did not pursue this feminist agenda for the next several decades had much to do with the life-and-death political confrontation between the

---

[11]For more on gender and the Chinese revolution, see Christina Gilmartin, *Engendering the Chinese Revolution* (Berkeley: University of California Press, 1995).

Figure 1. *Working in the Rice Paddies*
The backbreaking labor of cultivating rice in flooded paddies provides the
staple crop of much of China, as it has for centuries.
© Michael S. Yamashita/CORBIS.

CCP and the GMD. Likewise, since Marxism-Leninism taught urban
revolution by the proletariat, the CCP did not in the 1920s pursue the
rural revolution that Mao had suggested.

By the early 1930s, what was left of the CCP was running a rural
soviet (an elected government council) in Jiangxi province in southeast China. Throughout this period, the CCP survived by pursuing
guerrilla warfare—harrying the GMD's (and later Japan's) superior
forces with the "mobile warfare" outlined by Mao in 1936 (see Mao's
interview with Edgar Snow, Document 11).

In 1934 Chiang Kai-shek's modern army and air force drove the
CCP out of Jiangxi and on what became known as the Long March—
a three-thousand-mile retreat into the rugged hills of northwest China.

The survivors who set up shop in 1936 in the dusty county town of Yan'an (see Map 2 and Figure 2), some two thousand miles northwest of Jiangxi in Shaanxi province, were or soon became Mao's men. The Japanese invasion of central China in July 1937 forced Chiang to acquiesce to a second "united front" with the CCP, but this alliance was brief and weakened by mutual mistrust. It mostly gave the CCP protection from further GMD attacks, while both endured the brutal invasion by the Japanese army. Chiang led the national war effort from the GMD stronghold in Chongqing, in the far west of Sichuan province.

CREATING A NEW CHINA

By 1940 the Communist party had successfully passed through a tribulation of biblical proportions. Japan had encroached on Chinese territory since 1895, but in July 1937 the Japanese launched a blitzkrieg that took most of China's rich northern and central provinces. By 1942, after the Japanese attacked Pearl Harbor, the world was embroiled in the Second World War. At the war's end in 1945, China and its two dominant political parties, the Communists and the Nationalists, had endured eight years of bitter fighting. Even so, after a short standoff, the CCP and GMD fell into another civil war.

While resisting Japan, the CCP and GMD had competed for the hearts and minds of China's patriotic urban classes and rural majority. In this atmosphere, Mao wrote "On New Democracy," his blueprint for a "new China" (see Document 2). Published in a CCP journal in Yan'an in January 1940, this essay was not at first CCP policy. Mao's final rise to undisputed authority in the CCP took place in the Yan'an Rectification Movement of 1942–44. The Rectification Movement was an intense study program that focused on CCP history, why the CCP had been crippled by the GMD and forced on the Long March, and why Mao had the right ideas to fulfill the revolutionary promises of the CCP. By the end of this movement, the rank and file of the CCP, as well as its leaders (people called *ganbu,* or cadres), had a clear and compelling set of goals and an understanding of themselves, their role in China's rejuvenation, and even China's role in the liberation of oppressed peoples worldwide. Mao publicly introduced these ideas in "On New Democracy."

Mao cut an impressive figure during these years. Edgar Snow interviewed him in 1936 (see Document 11), shortly after Mao and his colleagues had set up camp in the Yan'an area. Snow saw "a gaunt, rather

Figure 2. *Yan'an and Its Cave Dwellings*
After the Long March and during the anti-Japanese war, CCP leaders worked in this isolated county seat and lived in the traditional peasant cave dwellings carved into the rugged hillsides. (Photo, 1981)
© Lowell Georgia/CORBIS.

14

Lincolnesque figure, above average height for a Chinese, somewhat stooped, with a head of thick black hair grown very long, and with large, searching eyes, a high bridged nose and prominent cheekbones. My fleeting impression was of an intellectual face of great shrewdness" (see Figure 3).

During the next decade, Mao articulated and engineered popular support for what became known inside and outside China as the Yan'an Way. Key aspects of the Yan'an Way are represented in two of Mao's other writings. In "Talks at the Yan'an Conference on Literature and Art" (see Document 3), which he delivered in May 1942, Mao gave his model for intellectuals in the new order and his quintessential example of ideological remolding for everyone. Included in these talks was the mass line, Mao's synthesis of the experience of the CCP administrations in various base areas, or rural soviets. He articulated a coherent policy of populist reform, rent reduction, and honest administration. The mass line also gave Communist leaders and local organizers a common set of goals and a common language—Maoism— with which to analyze China's problems and propose solutions. Mao's brilliant summary of this became the 1943 Central Committee's directive on the mass line (see Document 4), in which "correct leadership comes from the masses and goes to the masses."

Although the discipline the mass line required of cadres may seem extreme to readers today, we must remember the dire straits of both China and the CCP during World War II. Extraordinary danger called for extraordinary efforts, even at the cost of personal liberty. But the mass line was more than a harsh reaction to harsh conditions. Tang Tsou has shown that the mass line was not primarily a means of class struggle but rather a way to connect the goals of the CCP with the interests of ordinary Chinese. Brantly Womack rightly notes that during the years leading up to the Yan'an period, Mao shifted from a position of harsh dictatorship over landlords and other enemies of the CCP to "a politics sensitive to mass opinion within a mobilizational framework."[12]

In Yan'an, Mao found a way to mobilize his followers and increasing numbers of ordinary Chinese to support the nationalist and reformist goals of the CCP. As he said in "On New Democracy," he put off the

---

[12]Tang Tsou, "Reflections on the Formation and Foundations of the Communist Party-State," in Tang Tsou, *The Cultural Revolution and Post-Mao Reforms: A Historical Perspective* (Chicago: University of Chicago Press, 1986), 269–71; Brantly Womack, "The Party and the People: Revolutionary and Post-Revolutionary Politics in China and Vietnam," *World Politics*, 39, no. 4 (July 1987): 479–507.

Figure 3. *Mao Zedong in 1937*
This is how Mao looked when Edgar Snow met him in the late 1930s.
© Bettmann/CORBIS.

radical goals of communism to some future date and instead addressed the political and psychological concerns of most Chinese: their need to reside in a China of which they could be proud and to have enough social peace to live and work. The policies that Mao and the CCP developed in the Yan'an period addressed these needs convincingly. In the five years from 1944, when the Japanese began to falter, to 1949, when the CCP took control of Mainland China, the CCP put the Yan'an Way into practice and defeated the GMD. Mao felt a connection with these historic changes. In his poem "Snow," most likely penned in 1944 or 1945, he put himself on a par with China's famous founding emperors (see Document 5).

While in Yan'an, Mao was most in synch with China's revolutions— providing a compelling model for national mobilization, a gradual shift to socialist organization, and a way to reconnect China's educated elite to a government committed both to care for the rural majority and to build the modern industry necessary for economic development. Document 12, a selection from Stuart Schram's biography of Mao, provides a sense of the immense practicality and political success of Mao's model for a new China in Yan'an. This model did not come without a fight and was not without its ugly side. Mao had competitors for leadership, and he was ruthless in suppressing them. The very organization of Mao's writings comes out of the rectification campaign to discredit his enemies and unify the party under his hand. Those who got in his way were, like the party literary critic Wang Shiwei, cut down and vilified.[13]

In fact, during the Chinese civil war (1945–49), the CCP returned to the harsh class warfare in rural China it had carried out in Jiangxi in the early 1930s. The party abandoned the moderate rent reduction campaigns of the Yan'an period and revived full-scale land reform, in conjunction with mass denunciations, and often executions, of landlords. People's courts were held in villages throughout China not only to prosecute common criminals but also to give peasants a chance to "struggle against" local landlords and to distribute the landlords' land to poor peasants. As the CCP took over larger areas of China in the late 1940s, land reform—or land to the tiller—was a basic organizational tool of the party administration.[14] It removed the enemy (landlords allied with the GMD), it built a constituency (farmers freed from

[13]Still a valuable study of the politics of Yan'an and the rise of Mao is Raymond F. Wylie, *The Emergence of Maoism: Mao Tse-tung, Ch'en Po-ta, and the Search for Chinese Theory, 1935–1945* (Stanford: Stanford University Press, 1980). David Apter and Tony Saich give a vivid sense of Mao's inspiration to cadres in Yan'an and after in *Revolutionary Discourse in Mao's Republic* (Cambridge: Harvard University Press, 1994); see Document 19 for a review of this book. Finally, the tragic case of Wang Shiwei still haunts the halls of power in China, as seen in Dai Qing's reportage on the case, *Wang Shiwei and "Wild Lilies": Rectification and Purges in the Chinese Communist Party, 1942–1944* (Armonk, N.Y.: M. E. Sharp, 1994).

[14]Saich, *Rise to Power*, 1185–382, provides documents on the CCP's land policy in the context of the 1945–49 civil war. A vivid account of what land reform meant is covered by the American journalist Jack Belden, *China Shakes the World* (New York: Harper & Brothers, 1949), and in the fictional account of a Chinese participant written after she left China, Yuan-tsung Chen, *The Dragon's Village: An Autobiographical Novel of Revolutionary China* (New York: Penguin, 1980).

tenancy and grateful to the party), and it trained administrators in the goals of "serve the people."

Victory in the Chinese civil war brought new challenges for the CCP and for Mao. Chiang Kai-shek and the GMD regime retreated to the island of Taiwan, off the southeast coast of China. National consolidation by the CCP in 1949 solved one of the burning problems of the past century: ending national humiliation and foreign meddling inside China. Indeed, for almost all Chinese, Mao's declaration in September 1949 on the eve of the founding of the People's Republic of China (PRC) still brings a flush of pride: "The Chinese people have stood up!" (see Document 6). On October 1, 1949, Mao officially declared the founding of the PRC (see Figure 4).

But having stood up, the people had to figure where to go and how. The model for a new China outlined in "On New Democracy" was substantially modified on the eve of CCP victory in Mao's "On the People's Democratic Dictatorship."[15] Whereas in 1940 Mao saw the long-range program of the CCP—socialism—as farther off and the short-term program of a democratic united front as more pressing, by the end of the bloody civil war he and the CCP had reversed their emphasis to focus more on the exercise of "dictatorship over reactionary classes."

With increased confidence and revived radicalism, the CCP faced the question of implementing socialism in China and promoting economic recovery and development. Mao and the CCP chose the Soviet model—the Stalinist version of Marxism-Leninism and its centralized planned economy. Scholars have long noted that Mao was not a puppet of Stalin or the Soviet Union, and as we shall see, China would make an acrimonious split with the USSR in 1960. But this should not blind us to the real nature of the political system the CCP instituted in Yan'an and the new PRC. It was based on the Soviet model of the 1930s; it was, and still is, a Stalinist system. Although there were fewer purges and show trials in China than in the Soviet Union in the 1930s, and although China's approach to its farmers was much more solicitous than the Soviet Union's, recent research has confirmed the profound similarities between these two major socialist states.[16]

The central parts of the Soviet model were party hegemony over the government and military (without any constitutional or legal

---

[15]Dated June 30, 1949, in *Selected Works of Mao Tse-tung* (Peking: Foreign Languages Press, 1969), 4:411–24.

[16]Saich, *Rise to Power;* Timothy Cheek and Tony Saich, eds., *New Perspectives on State Socialism in China* (Armonk, N.Y.: M. E. Sharpe, 1997).

Figure 4. *Mao Speaking on Tiananmen Gate*
Mao declared the founding of the People's Republic of China, atop Tiananmen Gate in central Beijing, on October 1, 1949.
© Bettmann/CORBIS.

restraints), a centralized state administration that oversaw the operations of an increasingly complex national planned economy, a huge network of mass organizations (such as state-run labor unions, a national women's association, and a youth league), a muscular military, and an omnipresent secret police system. Running through this entire edifice was an independent network of party organizations and cells in every institution, down to the local party office, village, and neighborhood. It was a most amazing and extensive web. It was intended to make good on the promises of national rejuvenation and social equity made by the CCP in the 1940s, and it began with the good intentions of the cadres nurtured on the Yan'an model of rectification study and hands-on experience in the land reform struggle.

This idealistic regime confronted severe challenges. China's economy was wasted by some fifteen years of total war since the Japanese invasion in 1937. The United States enforced an almost immediate embargo on the new PRC. The Korean War broke out within ten months of independence (in June 1950), pulling China into battlefield confrontation with the most powerful nation in the world. These pressures brought out the dark side of the Yan'an Way—the Stalinist purges and inquisitions of those with connections to America, the search for spies, and the demand to conform to party ideology. In Yan'an, Mao had curtailed such purges, and in the early 1950s public campaigns against landlords, corruption, and spies were similarly contained. But the model of using mass campaigns in which millions of people across China followed directions from the national press to attack a particular problem or "enemy" was set.

Mass campaigns were part of a modified Stalinist model in which administrative issues were moralized into political crimes that were then attacked by a national campaign with token negative examples.[17] The early campaigns weakened competing social forces—landlords, capitalists, and bureaucrats left over from the GMD administration— but soon the campaigns became more ideological. Earlier efforts at reeducating intellectuals exploded into harsh media attacks on two leading scholars in 1955. Hu Shi, the great liberal academic and reformer of the New Culture Movement, was denounced as a bourgeois lackey because he had chosen to move to Taiwan with the ousted GMD regime rather than remain on the mainland under CCP

---

[17]The classic study of the origins and application of this rectification model is Frederick C. Teiwes, *Politics and Purges in China: Rectification and the Decline of Party Norms, 1950–1965,* 2nd ed. (Armonk, N.Y.: M. E. Sharpe, 1993).

rule. Hu Feng, the disciple of China's literary revolutionary hero Lu Xun, was attacked for his failure to adhere to the requirements of Mao's "Talks at the Yan'an Conference on Literature and Art" (see Document 3). Both media campaigns were integrated into the mass campaign style of rectification, with study groups in every party cell and office, every workshop and village. Instead of removing or disarming obvious enemies, such campaigns began to "eat the revolution's own." Thus the mass campaign was an effective but dangerous double-edged sword in the armory of militant socialism. It would soon come back to cut Mao and the CCP.

UTOPIAN REVOLUTIONS: THE HUNDRED FLOWERS CAMPAIGN, GREAT LEAP FORWARD, AND CULTURAL REVOLUTION

In 1956 China, the Communist party, and Mao had much to celebrate. The country was more unified, respected, peaceful, productive, and prosperous than it had been in a century and a half. Yet there were shadows lurking, and within a year horrible problems would surface, leading to national purges, the massive famines of the Great Leap Forward, and the political cataclysm of the Cultural Revolution. China had run up against the limitations of the Soviet model. Domestically, there was a shocking rise in bureaucratism: The Yan'an spirit of "serve the people" seemed to decrease in direct proportion to the consolidation of party power over society. Party leaders pushed others around and helped themselves to the fruits of victory. Internationally, socialism was experiencing real problems and challenges. These issues set Mao off in a direction that ultimately proved disastrous for China.[18] One of the first steps was his seminal writing "On the Correct Handling of Contradictions among the People" (see Document 7). It is arguably Mao's most important contribution to the development of state socialism and Marxist-Leninist theory after 1949.[19]

[18]This history and its consequences for China's politics are vividly portrayed in Roderick MacFarquhar, *The Origins of the Cultural Revolution*, vols. 1–3 (New York: Columbia University Press, 1974–97), and Frederick C. Teiwes with Warren Sun, *China's Road to Disaster* (Armonk, N.Y.: M. E. Sharpe, 1999).

[19]The other most obvious candidate is Mao's 1955 speeches, which were combined to become "On the Ten Great Relationships," in *Selected Works of Mao Tse-tung*, vol. 5 (1977). See Stuart Schram, *The Thought of Mao Tse-tung* (Cambridge: Cambridge University Press, 1989), 102.

Figure 5. *Anshan Steel Mill—Mao's Pride*
Anshan Steel Mill was famous for its Maoist workers' constitution in the spirit
of the Great Leap Forward. This picture of Anshan shows the successful
development of some heavy industry by the late 1950s in China.
© Milepost 92½/CORBIS.

Mao presented "On the Correct Handling of Contradictions among
the People" as a keynote lecture on February 27, 1957, to an expanded
group of party leaders. It was then circulated inside party channels
at special study meetings throughout the spring. It was finally published in a drastically edited form in June. The whole affair pointed out
an ominous development in Chinese socialism—the alienation of
the supreme leader from his party, his people, and ultimately from
reality.[20] Mao's concerns about increasing bureaucratism at home

[20]The contentious history of Mao's "Contradictions" speech and his push for a new
rectification movement has been obscured by the party's withholding of Mao's writings
from this period. In the 1980s, Red Guard copies of Mao's unedited speeches and talks
from January to July 1957 confirmed the inner-party struggle that Roderick MacFar-

were heightened by troubles in the international socialist world. First came Soviet premier Nikita Khrushchev's denunciation of Stalin in 1956 and the ensuing de-Stalinization in the Soviet Union and Eastern Europe. Next came the uprisings against Soviet control in Poland and Hungary. Mao accurately saw the likelihood of similar uprisings in China against an increasingly arrogant CCP administration, but his efforts to prevent such uprisings backfired.

Mao's approach to address any major public problem was the mass campaign. In this case, the problem was bureaucratism, the heavy-handed administrative tyranny of the CCP. To combat bureaucratism, Mao declared in early 1956 a reform movement under the slogan "Let a hundred flowers bloom; let a hundred schools contend." The idea was intended to reengage the technical and literary elite that had been chastened by the 1955 campaigns against Hu Shi and Hu Feng. Mao needed this elite to contribute to the next stages of industrialization and socialist transformation of commercial and village life. He sincerely wanted "the masses" of intellectuals and professionals to criticize party errors and excesses. In this aim, he was badly out of step both with his CCP compatriots and with most of China's intellectuals. The rest of the party held to the Leninist notion that the public could hardly comment usefully on the work of socialist transformation because most of the public was of bad, or at best neutral, class background—being mostly peasants and bourgeoisie, rather than working class. The party administration had no interest in soliciting public criticism from these suspect groups, but Mao insisted. In February 1957, using his "Contradictions" speech as a centerpiece, he pushed for a formal rectification movement of the sort that had melded the CCP into a winning team in Yan'an in the 1940s. The party apparatus resisted, but Mao ultimately got his way. A major public criticism campaign was launched in May 1957 as the Hundred Flowers Campaign. What Mao got from the public, however, was not tactical criticism on how to improve the party but strategic attacks on how to get rid of one-party rule. There were calls for parliamentary democracy, multiparty secret-ballot elections, and an end to CCP dictatorship.[21] The party apparatus's fears had come true, and Mao was left with egg on his face in front of his senior colleagues.

---

quhar and others had suspected. The core texts are translated and analyzed in Roderick MacFarquhar, Timothy Cheek, and Eugene Wu, eds., *The Secret Speeches of Chairman Mao: From the Hundred Flowers to the Great Leap Forward* (Cambridge: Harvard Contemporary China Series, 1989).

[21] See Goldman, *Literary Dissent in Communist China*, and Hualing Nieh, *Literature of the Hundred Flowers*, 2 vols. (New York: Columbia University Press, 1981).

The mistake was covered up in public. The open debate and criticism was shut down after barely a month, and the vicious Anti-Rightist Campaign was launched, in which those who had answered Mao's call to criticize the party had their lives ruined. They were fired, sent to virtual prison camps in undeveloped areas, and forever carried the stigma of rightist (that is, someone who opposed the CCP). All this was signaled, once again, by Mao's "Contradictions" piece. On June 6, 1957, it was published in the *People's Daily,* the official newspaper of the CCP, in a heavily edited form that lacked most of his earlier calls for public criticism of the party and included a restrictive litmus test of party loyalty that was to be applied to all such "suggestions." To the Chinese public, not to mention Western scholars bound up in cold war anti-Communist ideology, the Hundred Flowers Campaign appeared to be a particularly cynical attempt to lure opposition out into the open and then cut it off at the knees. Mao's reckless attempt to forestall the intellectual abuses of Soviet Stalinism had achieved the reverse result: It had created the Chinese Stalinist terror.

Understanding the inner history of the Hundred Flowers Campaign is essential in assessing what Roderick MacFarquhar calls the "Yan'an Round Table." King Arthur had fallen out with his knights.[22] From this time forward, Mao distrusted his party, and party leaders felt increasingly terrorized by their charismatic leader. The net result was that Mao became more bizarre and unpredictable, and his bureaucrats became more stodgy and cautious. This combination led to disaster.

Having called upon the intellectuals but found them wanting, Mao next turned to China's farming masses. In 1958 he declared the Great Leap Forward to complete China's two core tasks—finishing the socialist revolution to reach communism (the classless society) and achieving full economic development by overtaking Britain in industrial output. With the technical elite and local party cadres cowed after the Anti-Rightist Campaign, there was no organizational or professional resistance to the unrealistic plans Mao and his emerging circle of sycophants came up with. Mao's August 1958 "Talks at the Beidaihe Conference" (see Document 8) gives a sense of his dizzying imagination: the creation of "people's communes" (pooling the land that farmers had gained in land reform only six years earlier into communal farms), a return to a barter economy, communal dining halls, airstrips in every county, and backyard steel furnaces.

---

[22]See MacFarquhar, *Origins,* 3:471. The detailed history of the Hundred Flowers Campaign is covered in MacFarquhar, *Origins,* vol. 1.

Mao's profound detachment from the realities of life in China is captured by the somewhat tainted memoirs of one of his personal physicians, Dr. Li Zhisui (see Document 14). Mao's sequestered life of autocratic power was proving once again Lord Acton's dictum: "Power tends to corrupt, and absolute power corrupts absolutely." Tragically, the powerful organizational web that the CCP had built in the early 1950s responded loyally to Mao's impossible plans. Local leaders competed in reporting inflated grain harvests, finally claiming that thousands of pounds of rice or wheat were grown per acre. The party center in Beijing praised them and then took more tax revenue on the presumption that there really was more grain. But there was not. There was less, much less, because labor had been diverted to ill-founded industrial projects, such as the backyard steel furnaces. The result was the worst famine of the twentieth century. According to middle estimates, some 20 million Chinese, mostly in rural areas, died between 1959 and 1961.[23] Mao and the CCP were clearly to blame for this disaster.

What is amazing about the debacle of the Great Leap Forward is that the CCP did not fall from power. There was no counterrevolution. This is testimony to the residual legitimacy that the CCP had as the liberators of the Chinese nation from foreign bullies and of the Chinese farmer from tenancy. It is also testimony to the efficacy of centralized propaganda and the harsh political repression of critics in the Anti-Rightist Campaign of 1957–59. Years later, many Chinese reported that they thought the famine occurred only in their county, just an isolated catastrophe. Still, it was more than control of the media that got the recently landed farmers to give up their private land to the new collectives, the people's communes. Most people simply believed in Mao. Land reform was reversed after less than a decade, and the new landlord was the party-state, in the form of local officials who had an interest in making sure all land came back to the "collective."

The legacy of the Great Leap Forward is extensive. Not only did it lead directly to famine and then to the infighting of the Cultural Revolution in 1966, but even today party officials have not extricated themselves from the economy in China. What looks to advocates of trade liberalization with China as increasing privatization of the economy is actually the devolution of party economic control from central

---

[23]See Teiwes and Sun, *China's Road to Disaster*, and Jasper Becker, *Hungry Ghosts* (New York: Henry Holt and Co., 1996).

authorities in Beijing and regional authorities in the provincial capitals to local authorities or, worse, local party strongmen.[24] This profound mixing of business and government in China dates from the Great Leap Forward.

This second major disaster further alienated Mao from the CCP. He had officially "retired" from day-to-day party administration before the Great Leap Forward, but now he really did leave the actual leadership—and the cleanup—to his official successor, Liu Shaoqi. During the early 1960s, the party bureaucracy under Liu tried to pull China from the brink of economic and social collapse through a retreat to moderate social policies (a partial return to private farming and a relaxation in cultural life) and strict Leninist discipline. By 1965 China's economy was on the road to recovery. Yet for Mao, this economic recovery had come at the cost of revolutionary purity. He saw "sprouts of capitalism" in these moderate reforms.

During these years, Mao also picked an acrimonious fight with the leadership of the Soviet Union over ideological orthodoxy. This squabble prompted the departure of all Soviet technical advisers from China in 1960. The two Communist parties suspended relations, although the countries maintained chilly diplomatic channels. Good relations did not return until the late 1980s. Mao brooded over his revolutionary disappointments and concluded that he was surrounded by turncoats, or revisionists. In his view, the party itself was the problem because it was promoting capitalism. It needed to be destroyed and rebuilt.

The final scene in Mao's search to "continue the revolution" through utopian campaigns became the Cultural Revolution in 1966. He tried one more time to redeem the party through revolutionary struggle. Having found China's farmers as incapable as its intellectuals in creating the right sort of society, Mao turned to the youth. With his characteristic tactical brilliance, Mao maneuvered behind the scenes to set up a complex chess game that would bring down his perceived enemies (the party elite, including Liu Shaoqi). When he was ready, he set the game in motion with an attack on academic intellectuals that quickly flared into a full-scale assault on the Beijing party leadership. Mao then called on youth radicals to "Bombard the Head-

---

[24]Dorothy Solinger, "Urban Entrepreneurs and the State: The Merger of State and Society," in *State and Society in China: The Consequences of Reform,* ed. Arthur Lewis Rosenbaum (Boulder, Colo.: Westview Press, 1992).

quarters" (see Document 10). This gave supreme support to student attacks on all elders — teachers, parents, and party officials.[25]

The results in 1966 and 1967 were public terror at the hands of teenage gangs called Red Guards. They were officially sanctioned by Mao, paraded several times before him in Tiananmen Square in central Beijing, and wrought havoc around the country (see Figure 6). Rae Yang's memoir (see Document 15) gives us a glimpse of the madness of the Cultural Revolution. The Red Guards' bible was *Quotations from Chairman Mao Zedong*, commonly known as the "Little Red Book" (see Document 10). Lin Biao, the head of the military, had edited it a few years earlier as a simplified study guide for soldiers. Hundreds of millions of copies of the book were distributed, as the country teetered on the brink of chaos.

The excesses of the Cultural Revolution were not the end of Mao. He lived for another decade. During those years, both he and the CCP reeled in the excesses of the Red Guards by bringing in the People's Liberation Army (PLA) to restore order in China's schools, offices, and villages. Red Guard gangs were broken up in the "rustification movement," in which individual teenagers were "sent down" to villages across China to "learn from the peasants." Since Mao had purged his designated successor, Liu Shaoqi, in 1967, he chose a new successor, Lin Biao. Yet Mao soon came to suspect his new successor as well, and in 1971 Lin Biao died in a plane crash while trying to elude Mao's secret police. During this time, various factions in the CCP competed to get Mao's attention and approval. The two biggest groups were those around his wife, Jiang Qing, who were later known as the Gang of Four, and those around Prime Minister Zhou Enlai. In fact, Mao supported both groups some of the time.

Mao had one more revolution up his sleeve. After two decades of cold war antagonism (see Document 9), Mao surprised everyone by restoring diplomatic relations with the United States. First with secret negotiations with U.S. secretary of state Henry Kissinger in 1971 and then with President Richard Nixon's visit to China in 1972, Mao

---

[25]The best documentary collection is Michael Schoenhals, ed., *China's Cultural Revolution, 1966–1969: Not a Dinner Party* (Armonk, N.Y.: M. E. Sharpe, 1996). Personal accounts of life during these years are a cottage industry in America. Among the most engaging for students are the memoirs by Gao Yuan, *Born Red* (Stanford: Stanford University Press, 1985), and Rae Yang, *Spider Eaters* (Berkeley: University of California Press, 1997). Also see Anita Chan, *Children of Mao: Personality Development and Political Activism in the Red Guard Generation* (Seattle: University of Washington Press, 1985).

Figure 6. *Red Guards Parading during the Cultural Revolution*
Red Guard student activists parade in Tiananmen Square, Beijing, in 1967.
They carry a poster of Mao in front, labeled "Long Live Chairman Mao." The
big set of placards reads "Hold High the Great Red Banner of Mao Zedong
Thought," and the smaller banner toward the back reads "Long Live the Great
Proletarian Cultural Revolution." Such mass rallies were common at the begin-
ning of the Cultural Revolution in 1966 and 1967.
© Bettmann/CORBIS.

reversed more than twenty years of animosity between the two countries. With U.S. support, the PRC took the "China seat" on the UN
Security Council, displacing the GMD regime of Chiang Kai-shek
(which occupied Taiwan). The United States and China finally normal-
ized relations and exchanged ambassadors in 1979. Normal diplomatic
relations between China and the United States, despite our various
tensions, are the result of Mao's last, dramatic act of pragmatic leadership of China.

When Mao died in September 1976, China was struck with grief. It
was a mixed grief, combining the memory of both Mao's great contri-

butions to the establishment of the nation, the application of socialist goals, and the realization of economic progress, and his cruel persecution of intellectuals, his abandonment of the peasants, his ruinous mass campaigns, and the arbitrary dictatorship of his later years. When Mao was born, China was a failing, backward empire. When he died, it was a respected (or feared) nation on the world scene—a nuclear power with a permanent seat on the UN Security Council and able to feed its people and run a modern economy. The price of this success had been high, and the credit for it belongs to thousands of hardworking Chinese, not just to Mao. But Mao came to represent the noble goals, the grand achievements, and the terrible failures of China in the twentieth century.

EXPERIENCING MAO'S REVOLUTIONS

The readings in part two offer some glimpses of how people inside and outside China experienced Mao. From early on, it was clear that Mao captured people's imaginations as a hero and potential savior. In Document 11, Edgar Snow, his first biographer, records popular stories about Mao as early as 1936. "Mao had the reputation of a charmed life," writes Snow. "He had been repeatedly pronounced dead by his enemies, only to return to the news columns a few days later, as active as ever." Writing before Mao's final rise to power, Snow concludes, "There would never be any one 'savior' of China, yet undeniably one felt a certain force of destiny in Mao."

In the 1930s, Mao was engaged in constant warfare—defending his rural "people's soviet" from local bandits or from the government troops of the GMD, and later resisting the Japanese invasion of China. He was significant mostly to rural people who had to deal with harsh landlords, violent bandits, and corrupt local officials. The Red Army that Mao came to lead, as described in Document 11, shows how effective Mao, Zhu De, and other Communist leaders were in addressing these rural problems.

Stuart Schram's biography (see Document 12) gives us a sense of what it was like to apply Mao's ideas at Yan'an, the Communist capital during World War II. Mao's Thought was a big part of his attraction to many Chinese. He was not just an inspiring figure and a successful general. His way of explaining life made sense to people. Nick Knight offers a good description of how Mao's philosophy applied to broader issues and why it attracted so many Chinese (see Document 13). Mao made Marxism-Leninism—a European ideology developed by Karl Marx, a German, and applied by Lenin, a Russian—into a Chinese system of thought. That is, Mao Sinified Marxism. Coming into the 1950s, a Chinese farmer would be grateful for the land given to him by CCP land reform; a Chinese worker would be happy with better working conditions; a student would be thrilled at the chance to serve China through the CCP, which had brought these good things to pass; and all would be proud that Mao and the CCP had made China stand up as an independent nation at peace for the first time in a century. It is no wonder that Mao's "Sinified Marxism" seemed the right way to ensure China's future growth and development.

Yet the promise was not fulfilled, at least not without a terrible price. By the end of the 1950s, only ten years after their inspiring national victory, Mao and the CCP had developed a police state and caused the biggest famine in China's history. Today researchers can analyze why this happened, but at the time most people could not understand why things were going wrong. The memoir by Li Zhisui, one of Mao's personal doctors (see Document 14), gives one reason: Power corrupts. Mao was a superstar by the 1950s and lost touch not only with the common people but with his colleagues. Yet the more distant Mao became from his people and his peers, the harder it seemed to blame him for what was going wrong. Despite the suffering of the Great Leap Forward, Mao was idolized (see Figure 7). When he unleashed the Cultural Revolution in 1966, hardly a soul thought, or dared, to contradict anything he said.

At the peak of his power, Mao had drifted away from Chinese reality. Whereas his 1927 "Report on the Peasant Movement in Hunan" was a detailed analysis of meticulous field research, the basic Cultural Revolution reading, *Quotations from Chairman Mao Zedong,* provided soothingly simplistic sound bites that were used in often superstitious ways by a confused and troubled public. Red Guards reported a mystical sense of union just by chanting "Long live Chairman Mao," and more recent Chinese scholars note that applying a single Mao sentence to some practical problem continued the superstitions of folk religion (see Documents 15 and 18).

Although Mao's 1957 speech "On the Correct Handling of Contradictions among the People" had failed to create a legitimate arena for public criticism under state socialism (largely due to his own impetuosity), ironically the social disaster of the Cultural Revolution sowed the seeds of democracy in China. A generation of youths learned that it was indeed "right to rebel" (Mao's phrase from his 1927 Hunan

主席是我们心中永远不落的红太阳

Figure 7. *Father Mao Poster from the Cultural Revolution Period*
This is one of literally thousands of colorful propaganda posters idolizing Mao.
He is pictured as the patriarch overseeing the revolutionary masses, who
clutch copies of his red *Quotations from Chairman Mao.* The poster's title, at
the bottom, reads "Chairman Mao Is the Never-Setting Red Sun in Our
Hearts." This particular print is from 1978.
© Ric Ergenbright/CORBIS.

report, which was lionized out of context throughout the Cultural Revolution), but they also learned that the Communist party could be
wrong. Rae Yang, who was one of the youths "sent down" to the
villages during the Cultural Revolution (see Document 15), later reflected on her experience: "Now I agree with Chairman Mao that class
struggle continues to exist in China under socialist conditions—but
not between landlords and poor peasants or capitalists and workers. It

goes on between the Communist Party officials and the ordinary Chinese people!"[26]

In the late 1960s, millions of educated youths were "sent down" to the rural areas ostensibly to "learn from the masses" but actually to disband the Red Guard gangs in the cities. It took nearly ten years for some of these youths to make it back to China's post-Mao universities. Today that cohort of intellectuals and leaders, now in their fifties, has a profound understanding of and link to rural China and a deep-seated distrust of authority.[27] Mao unintentionally provided a solution to two long-standing problems in twentieth-century China—the alienation of the urban elite from China's rural majority and the burden of authoritarian government.

Empowering intellectual independence and reconnecting China's elite with the countryside were not the party's goals in the Cultural Revolution. The CCP was simply trying to survive this final assault by its vengeful founder. When Mao died, he left a wounded institution behind. But it was not mortally wounded, as its survival into the twenty-first century demonstrates. We can see Mao's legacy in the CCP's survival and in the problems it faces. Deng Xiaoping, who was part of Zhou Enlai's faction against Jiang Qing and the Gang of Four, became China's paramount leader after Mao. Deng built his popular support by analyzing and addressing the needs of China's rural majority. He explicitly saw himself as a Maoist in this regard, even though his actual success was based on dismantling Mao's beloved people's communes and restoring the land to those who had originally been awarded it under land reform.

The CCP today uses Mao's united front policies from "On New Democracy" to organize a more relaxed control over public life. In 1988 Li Zehou, a noted philosopher and cultural commentator, invoked "New Democracy" as the right balance of capitalist forces and socialist goals. A decade later, in 1999, senior party historian and CCP establishment spokesman Hu Sheng revisited "On New Democracy" in the flagship journal of the Chinese Academy of Social Sciences, *Chinese Social Sciences*. Hu came to essentially the same conclusion that Li had a decade earlier: New Democracy made the proper use of capitalist energy (farmers and merchants) for socialist goals (a strong and prosperous China) because it focused on the development of productive forces (economic development).[28] As Mao's most senior Western

---

[26] Rae Yang, *Spider Eaters*, 163.
[27] Lisa Rafel, *Other Modernities: Gendered Yearnings in China after Socialism* (Berkeley: University of California Press, 1999).
[28] The interesting thing to note about these two views, aside from their continuity

expert, Stuart Schram, has noted, Mao's core writings from 1935 to 1965 served "as a vehicle of Westernization" that helped China become a modern nation.[29]

Finally, Mao's search for a mechanism for legitimate political opposition under state socialism—the heart of his "On the Correct Handling of Contradictions among the People"—enlivens party leaders today. In 2001 party chairman Jiang Zemin tried to get businessmen to join the CCP.[30] Quite possibly, such an expanded and liberal version of the CCP will become one building block of "democracy with Chinese characteristics."

Mao's legacy is contested. In 1981 the CCP Central Committee passed a resolution on party history and the role of Mao (see Document 16). In this resolution, the Central Committee tried to enforce a formula of 70 percent successes and 30 percent failures in Mao's life and revolutionary leadership. But this resolution has been challenged by writers such as Li Zhisui, who published a denunciation of Mao as an evil emperor in 1994 (see Document 14). This book has been translated into Chinese and circulates widely (albeit illegally) in China today.[31] In Document 17, Jeffrey Wasserstrom provides a sense of how scholars approach Mao today in a review of Li's book and a study by David Apter and Tony Saich of the charismatic authority of Mao in the minds of party cadres. Wasserstrom identifies the major trend in recent studies on Mao and Chinese history: a shift in focus from Mao the man, as representative of a single Chinese socialist revolution, to Mao the image, in the hearts and minds of hundreds of millions of Chinese. We hear some of those unofficial voices from China today in Document 18.

Politically, Mao's legacy is ambiguous. Political scientist Frederick Teiwes notes the reason: "Mao degenerated from pragmatic visionary

---

across the "great divide" of the 1989 Tiananmen protests and repression, is that Li Zehou is considered by Beijing authorities to be a "dissident," while Hu Sheng is an acknowledged establishmentarian. See *Li Zehou wenji* (Works of Li Zehou) (Taipei: Sanmin shuju, 1996), 345, reprinted from *Renmin ribao*, 7 Apr. 1988, and Hu Sheng, "Mao Zedong de Xin minzhuzhuyi lun zai pingjia" (Assessment of Mao's "On New Democracy"), *Zhongguo shehui kexue* (Chinese Social Sciences), 3 (10 May 1999): 4–19.

[29]Schram, *The Thought of Mao Tse-tung*, 192.

[30]Special report on China, *Economist*, 30 June 2001, 9, 21–23; Craig Smith, "Workers of the World, Invest!" *New York Times*, 19 Aug. 2001, WK3.

[31]It is controversial not only with CCP authorities but also with other scholars who feel that the Chinese edition (as well as the English) is distorted. See Geremie R. Barmé, *Shades of Mao: The Posthumous Cult of the Great Leader* (Armonk, N.Y.: M. E. Sharpe, 1996), 53, 72. Li's book has been criticized as bad history by some serious Western scholars, notably Frederick C. Teiwes, "Seeking the Historical Mao," *The China Quarterly*, Mar. 1996.

Figure 8. *Mao Lighter*
One of the popular uses of Mao's memory and image in China today is this
striking red cigarette lighter that is still widely available in urban China in the
early 2000s. The Chinese characters above Mao's head read "The Red Sun,"
one of the accolades given to the Chairman during the Cultural Revolution.
When opened to light, the red light at the top-center flashes and the tune "The
East Is Red" (also of Cultural Revolution vintage) plays. This is a prime
example of the commercialization and nostalgia for Mao among a generation
that did not actually experience the Cultural Revolution.
Jack Hayes; used with permission.

to paranoid ideologue." On the one hand, Mao's great charisma and
authority have stood the test of time, making his leadership style popular even today. On the other hand, the leaders who worked with Mao,
and now a new generation that has learned those lessons secondhand,
desire "normal politics" without the chaotic mass campaigns and terri-
fying purges that Mao promoted. Even so, they have inherited political
practices that still cause problems today. The CCP still carries a "them
and us" attitude as a revolutionary party that feels it cannot trust the
ordinary citizen, and it is unwilling to limit its power by establishing a
viable legal system. Furthermore, the political methods of Mao continue to haunt the CCP, as is evident in their faith in a now discredited
ideology, their reliance on a "core leader," the practice of lower-level
leaders clambering to endorse that top leader's "thought," and a mass

Figure 9. *Mao in Temple*
Another use of Mao in contemporary China has been to incorporate him into the vibrant revival of popular religion. Here a Cultural Revolution–period portrait of Mao appears in a local temple in Lahsa, Tibet (1983). Mao clearly remains a potent figure in the popular imagination, but not necessarily for the idealistic policies he advocated or the terrible tragedies he precipitated. What meanings Chinese of the twenty-first century will draw from Mao is an open question, but they are likely to make use of his legacy in some fashion.
© Bettmann/CORBIS.

campaign approach to carrying out policy. "Until these practices are overcome," Teiwes concludes, "it is unlikely that the most important aspect of 'stability and unity'—that between the regime and society— can be achieved."[32]

In all, Mao's revolutions—at first ahead of his contemporaries, then as the best expression of them, and finally as unreasonable dictator—contributed to China's broader revolutions in nationalism,

[32]Frederick C. Teiwes, "Politics at the 'Core': The Political Circumstances of Mao Zedong, Deng Xiaoping and Jiang Zemin," *China Information*, 15, no. 1 (2001): 1–66.

socialism, and economic development. Like other efforts to revolutionize China, Mao's efforts and those of the CCP ran up against the realities of village and urban life that had confronted emperors and mandarins for a thousand years: the importance of family, the thick web of local relationships in villages and urban neighborhoods that defends against efforts by the state to get people to do what it wants, the resilience of "popular religion" that remains out of state control (most recently the Falungong group), and the need to employ and care for a huge and growing population.[33] The clash between these revolutionary impulses and continuing realities has produced not revolutionary change in China today, but evolution—a gradual development of life and thought.

Thus Mao's final legacy is not a simple matter. He will not be forgotten, but he also will no longer likely serve as the charismatic font of all wisdom. Mao has crossed the Taiwan Strait to the home of his Nationalist party rivals, establishing new and developing links between Mainland China and Taiwan. An establishment scholar, Professor Jiang Yihua at Fudan University in Shanghai, edited a volume of Mao's writings that was published in Taiwan in 1994.[34] Radical scholars such as Cui Zhiyuan may call upon Mao's mass campaigns as a model of "extensive democracy" to combat the corruption in China today,[35] and ordinary people may see Mao as a commercial image or as another deity in the pantheon of China's enduring folk religion (see Figures 8 and 9 and Document 18). In the end, some of his contributions will be honored and continued, and others will be discarded. Certainly, his fierce nationalism and faith in China's ordinary people, his focus on the farmers, and his fearless willingness to experiment will likely find a ready audience in the China to come.

[33] On these daily realities and their continuity through centuries of political change, see Lloyd Eastman, *Family, Fields, and Ancestors: Constancy and Change in China's Social and Economic History* (New York: Oxford University Press, 1988); James L. Watson, *Class and Social Stratification in Post-Mao China* (Cambridge: Cambridge University Press, 1984); and David Ownby, "A History for Falungong," *Nova Religio,* Vol. 5, No. 2 (2002, forthcoming).

[34] *Mao Zedong zhuzuo xuan* (Selection of Writings by Mao Zedong), ed. Jiang Yihua (Taipei: Taiwan Shangwu yinshuguan, 1994).

[35] Cui Zhiyuan, quoted in Gloria Davies, ed., *Voicing Concerns: Contemporary Chinese Critical Inquiry* (Lanham, Md.: Rowman & Littlefield, 2001), 52.

PART ONE

Mao Documents

A NOTE ABOUT THE TEXTS

The introduction provides the historical context of Mao Zedong's writings and life. However, it is also the task of historians to assess these texts through source criticism.[1] Historians need to know how the words we read in a document come to us so that we can assess those words. The most obvious point to note here is that the texts in this reader are translations. Mao wrote in modern Chinese; he knew no foreign languages. The translations used here are the best available, by the finest scholars in the field or by the official Beijing translation committee of Mao's *Selected Works*. However, in the case of Mao's writings, evaluation of these documents is complicated by his stature as the "Great Helmsman," the "Savior of the Chinese People," and the author of "Marxism-Leninism Mao Zedong Thought," which is the official ideology of the CCP and thus of China. Thus the problems of the textual transmission and editing of Mao's writings most nearly resemble those of theological texts, such as biblical writings. Fortunately, recent scholarship, especially in China, helps us to know which sort of Mao text we are reading.[2]

Since the advent of "Mao Zedong Thought" in the early 1940s, Mao's writings have been published in China in three kinds of collections. These may be distinguished by the intent of their editors as "collective wisdom" editions, "genius" editions, and "historicist"

---

[1] John Tosh, *The Pursuit of History*, 3rd ed. (London: Longman, 2000).

[2] For much more detail on the source criticism of Mao's texts, see Timothy Cheek, "Textually Speaking: An Assessment of Newly Available Mao Texts," in *The Secret Speeches of Chairman Mao: From the Hundred Flowers to the Great Leap Forward*, ed. Roderick MacFarquhar, Timothy Cheek, and Eugene Wu (Cambridge: Harvard Contemporary China Series, 1989), 75–103.

editions.[3] First, beginning in 1944, volumes of *Selected Works of Mao Zedong* began to appear by the order of various high-level CCP institutions. They were edited by committee according to a "collective wisdom" criterion: the idea that Mao "represented" the summation of Sinified Marxism-Leninism and thus should reflect the consensus of the party leadership. That this editing relied most heavily on Mao's acquiescence and that the process was highly distorted as early as 1960 did not weaken, in the minds of the editors, the attempt to make official Mao works a "collective" enterprise. Indeed, both Mao himself and advisers from the Soviet Union were active in this process during the early 1950s. Second, during the Cultural Revolution and particularly at the height of the Red Guard movement in 1967, Mao's writings were published by a confusing array of unnamed editors based on the idea that the Chairman was a lone genius and thereby not subject to revision by any collective leadership, least of all by a party riddled with "capitalist roaders." Finally, since Mao's death, party historiographers have published both restricted circulation and publicly available collections of his writings that reflect in varying degrees a "historicist" urge to understand the past as it really was and to place Mao and his writings more firmly in a historical context.

Most of the texts in the first part of this reader are taken from official, or "collective wisdom," editions. I have selected these versions, even though they have been more or less heavily edited from the original, because they are the versions that were studied by hundreds of millions of Chinese as the authoritative word of Mao and the doctrine of the CCP.[4] Some of the shorter extracts come from "historicist" or "genius" editions to give the reader a sense of the difference of these significant but less widely read versions of Mao's writings. The introduction to each document notes its precise source and nature.

The reader of any text that comes from one of Mao's official *Selected Works* needs to keep two points in mind. First, the texts have been freely edited to conform to what the editors considered to be the needs of changing circumstances. That is, in contravention of scholarly standards of accuracy in reprinting historical documents, the texts have been altered. Many well-known studies have documented this

---

[3]These are my categories and not Chinese ones, although they correspond (in order) to the Chinese categories of official (published by the Mao committee), unauthorized, and reference (*yanjiu*, or "research") in the publication of Mao's writings.
[4]As many as 236 million copies of the first four volumes alone were published during Mao's life, according to Michael Y. M. Kau and John Leung, eds., *The Writings of Mao Zedong, 1949–1956* (Armonk, N.Y.: M. E. Sharpe, 1986), 1:xxvi.

phenomenon.[5] For example, Mao's 1957 speech "On the Correct Handling of Contradictions among the People" was massively edited, but I have used the edited version as it appears in volume 5 of his *Selected Works* because that was the version that was widely studied (see Document 7). Among the more common changes in these texts are the deletion of positive comments about the GMD, earthy language, references to policies that did not work, and specific details. Scholars who want to research Mao's role and intentions in specific contexts must seek out the original versions, which are widely available.[6] For those wishing to contemplate the *impact* of the Chairman's writings on the Chinese reading public, the *Selected Works* versions are more appropriate.

Second, the reader should be aware that the official editions of Mao's writings published before his death in 1976 are considered by the CCP leadership to represent "the crystallization of collected wisdom in the CCP."[7] This has been the case since "Mao Zedong Thought" was officially designated as the guiding thought of the CCP in the June 1945 constitution passed at the Seventh Party Congress in Yan'an. This is the trade-off for letting Mao speak for the party: The party can check what he said in print.[8]

[5]See Jerome Chen, *Mao Papers: Anthology and Bibliography* (London: Oxford University Press, 1970), introduction.

[6]Takeuchi Minoru, *Mao Zedongji* (Collected Writings of Mao Zedong), rev. ed., 10 vols. (Tokyo: Hokobosha, 1983), and *Mao Zedong ji bujuan* (Supplements to Collected Writings of Mao Zedong), 10 vols. (Tokyo: Sososha, 1983–86). The tenth volume of *bujuan,* which is unnumbered, contains a chronology of Mao's writings. These sets constitute the most reliable collections of Mao's writings in Chinese, along with *Mao Zedong wenji* (Collected Writings of Mao Zedong), 8 vols. (Beijing: Renmin chubanshe, 1993–99), which goes only through 1957.

The standard references for translations of Mao's collected works are Stuart R. Schram and Nancy J. Hodes, eds., *Mao's Road to Power: Revolutionary Writings, 1912–1949,* planned 10 vols. (Armonk, N.Y.: M. E. Sharpe, 1992–2003), of which five volumes have appeared by 2001, and for September 1949 to December 1957, Michael Y. M. Kau and John K. Leung, eds., *The Writings of Mao Zedong, 1949–1976,* 2 vols. (Armonk, N.Y.: M. E. Sharpe, 1986, 1992). Unfortunately, the Kau and Leung volumes will go no further. The official, or "collected wisdom," edition of *Selected Works of Mao Zedong* in Chinese and English, as edited by the CCP and published in Beijing, is widely available, with a corrected edition released for Mao's centenary in 1993. Finally, full texts of the official English version of *Selected Works* are available on the Web at <http://www.marx2mao.org/Mao/Index.html>.

[7]A classic example of this claim can be found in paragraph 28 of the Central Committee's June 1981 resolution, "On Questions of Party History." For the English version of this resolution, see *Beijing Review,* 6 July 1981, 10–39; paragraph 28 also appears in Helmut Martin, *Cult & Canon: The Origins and Development of State Maoism* (Armonk, N.Y.: M. E. Sharpe, 1982), 213.

[8]See, for example, Martin, *Cult & Canon,* and Joshua A. Fogel, *Ai Ssu-chi's Contribution to the Development of Chinese Marxism* (Cambridge: Harvard Council on East Asian Studies, 1987).

The current guide inside China for permissible interpretations of CCP history and Mao's writings is "Resolution on Some Questions on Party History Since the Founding of the Nation," passed by the Central Committee on June 27, 1981 (see Document 16).[9] Perhaps the best preliminary metaphor for the issues surrounding historicist party writings in the 1980s, including those on Mao's works, is that of academic theology in the Christian and Jewish traditions, where "scientific" linguistic and historical analyses seek to contribute to a living faith.

Finally, an assessment of Mao's writings would not be complete without at least a reference to the continuing debate in the West over the interpretations of these texts. Some of the documents in this book reflect these issues (see Documents 13 and 17). The reader should, however, be aware of a few issues in order to mine Mao's texts independently. Stuart Schram's basic text, *The Political Thought of Mao Tse-tung*, and his collection of alternative Cultural Revolution "genius" editions of Mao's works, *Mao Zedong Unrehearsed*, provide ample background on these issues, as does Roderick MacFarquhar and company's *Secret Speeches of Chairman Mao*. In addition, Schram has produced a review of the literature, "Mao Studies: Retrospect and Prospect," which raises problems of interpretation.[10] More provocatively, Nick Knight challenges the reader to confront the problems of relativism and unconscious assumptions in analyzing Mao's texts.[11]

In the selections, footnotes original to the document use footnote symbols (*, †, ‡, etc.); footnotes added by the editor of this volume are numbered.

[9]*Beijing Review*, 6 July 1981, 10–39; Martin, *Cult & Canon*, 180–231.

[10]Stuart Schram, *The Political Thought of Mao Tse-tung*, rev. ed. (New York: Praeger, 1969); Stuart Schram, *Mao Zedong*, rev. ed. (Harmondsworth: Penguin, 1967); Stuart Schram, "Mao Studies: Retrospect and Prospect," *China Quarterly*, 97 (1984): 95–125. See also, Brantly Womack's thoughtful review essay on the first five volumes of Schram and Hodes's *Mao's Road to Power:* "Mao before Maoism," *The Chinese Journal*, no. 46 (July 2001): 95–117.

[11]Nick Knight, "Mao and History: Who Judges and How?" *Australian Journal of Chinese Affairs*, no. 13 (January 1985): 121–36. Knight makes a further contribution in "Mao Zedong: Ten Years After," *Australian Journal of Chinese Affairs*, no. 16 (July 1986): special issue.

1

Report on the Peasant Movement in Hunan

February 1927

This is one of Mao's most famous essays. It sets out to describe the uprising among poor farmers in the counties outside Changsha, in the central province of Hunan, in the winter of 1926–27. The full text of this report was first published in a local CCP journal in March and April 1927. Mao's intended audience was his fellow revolutionaries in the CCP and in the revolutionary wing of the GMD, or Nationalist Party.

The essay is organized into three broad topics: (1) an enthusiastic description and defense of the rural violence that had engulfed the area (and parts of neighboring provinces); (2) a class analysis of the leadership of this violence; and (3) a fourteen-point account of the achievements of the peasant associations that had sprung up to replace the decimated traditional power structure. Throughout the essay, Mao declares that this is the real Chinese revolution, not the urban revolution promised by European Marxism. Mao's genius was to locate the fundamental battleground of this class struggle not at the national or provincial level, but in local society. Mao especially claims that this rural insurrection fulfills the promise of Sun Yat-sen's 1911 Revolution, which had been subverted by militarists.

The essay includes one of Mao's most famous maxims: "Revolution is not a dinner party" (it is phrased slightly differently in this translation). This famous defense of the necessity and appropriateness of revolutionary violence, particularly summary executions of unpopular landlords, marked Mao as a serious and, among the GMD, hated revolutionary. He never looked back.

"Hunan nongmin yundong kaocha baogao," in *Mao Zedong ji*, ed. Takeuchi Minoru (Tokyo: Hokobosha, 1970), 1:207–49, which is taken from the 1944 and 1947 Chinese editions of Mao's *Selected Works*. Translation from Stuart R. Schram and Nancy J. Hodes, eds., *Mao's Road to Power: Revolutionary Writings, 1912–1949: Volume II: National Revolution and Social Revolution, December 1920–June 1927* (Armonk, N.Y.: M. E. Sharpe, 1994), 2:429–64. I have edited the text very lightly to minimize technical terms. I have not included Schram's extensive notations of textual variations between this original and the post-1949 editions of Mao's *Selected Works*.

I. RURAL REVOLUTION

1. The Importance of the Peasant Problem

During my recent visit, I made a first-hand investigation of the five counties of Xiangtan, Xiangxiang, Hengshan, Liling, and Changsha. In the thirty-two days from January 4 to February 5, I called together fact-finding conferences in villages and county towns, which were attended by experienced peasants and by comrades in the peasant movement, and I listened attentively to their reports and collected a great deal of material. Many of the arguments of the peasant movement were the exact opposite of what I had heard from the gentry class in Hankou and Changsha. I saw and heard many strange things of which I had hitherto been unaware. I believe that the same is also true of every province in all of China. Consequently, all criticisms directed against the peasant movement must be speedily set right, and the various erroneous measures adopted by the revolutionary authorities concerning the peasant movement must be speedily changed. Only thus can the future of the revolution be benefited. For the present upsurge of the peasant movement is a colossal event. In a very short time, several hundred million peasants in China's central, southern, and northern provinces will rise like a fierce wind or tempest, a force so swift and violent that no power, however great, will be able to suppress it. They will break through all the trammels that bind them and rush forward along the road to liberation. They will, in the end, send all the imperialists, warlords, corrupt officials, local bullies, and bad gentry to their graves. All revolutionary parties and all revolutionary comrades will stand before them to be tested, to be accepted or rejected as they decide. To march at their head and lead them? To stand behind them, gesticulating and criticizing them? Or to stand opposite them and oppose them? Every Chinese is free to choose among the three, but by the force of circumstances you are fated to make the choice quickly. Here I have written up my investigations and opinions in several sections, for the reference of revolutionary comrades.

2. Get Organized

The peasant movement in Hunan, so far as it concerns the counties in the central and southern parts of the province, where the movement is already developed, can be roughly divided into two periods. The first, from January to September of last year, was one of organization.

Within this period, January to June was a time of secret [activity], and July to September, when the revolutionary army was driving out Zhao,[1] an open time. During this period, the membership of the peasant associations did not exceed 300,000 to 400,000, and the masses directly under their command numbered little more than a million; there was as yet hardly any struggle in the rural areas, and consequently there was very little criticism of the associations in other circles. Because its members served as guides, scouts, and porters, even some of the officers had a good word to say for the peasant associations. The second period, from last October to January of this year, was one of revolution. The membership of the associations jumped to 2 million and the masses directly under their command increased to 10 million. (The peasants generally enter only one name for the whole family on joining a peasant association; therefore a membership of 2 million means a mass following of 10 million.) Almost half the peasants in Hunan are now organized. In counties like Xiangtan, Xiangxiang, Liuyang, Changsha, Liling, Ningxiang, Pingjiang, Xiangyin, Hengshan, Hengyang, Laiyang, Chenxian, and Anhua, nearly all the peasants have gone into the peasant associations or have come under their command. It was on the strength of their extensive organization that the peasants went into action and within four months brought about a great revolution in the countryside, a revolution without parallel in history.

3. Down with the Local Bullies and Bad Gentry! All Power to the Peasant Associations!

Now that the peasants have got themselves organized, they are beginning to take action. The main targets of their attack are the local bullies, the bad gentry, and the lawless landlords, but in passing they also hit out against patriarchal ideas and institutions of all kinds, against the corrupt officials in the cities, and against bad practices and customs in the rural areas. In force and momentum the attack is quite simply tempestuous; those who submit to it survive, and those who resist perish. As a result, the privileges the feudal landlords have enjoyed for thousands of years are being shattered to pieces. Their dignity and prestige are being completely swept away. With the collapse of the power of the gentry, the peasant associations have now become the sole organs of authority, and "All power to the peasant

[1] *Zhao:* Zhao Hengti, governor of Hunan until March 1926

associations" has become a reality. Even trifling matters such as quarrels between husband and wife must be brought before the peasant association for settlement. Nothing can be settled in the absence of peasant association representatives. Whatever nonsense the people from the peasant association talk at meetings, that, too, is sacred. The association actually dictates everything in the countryside, all rural affairs, and quite literally, "whatever it says, goes." People outside the associations can only speak well of them and cannot say anything against them. The local bullies, bad gentry, and lawless landlords have completely lost their right to speak, and none of them dares even mutter dissent. Faced by the intimidating force of the peasant associations, the top local bullies and bad gentry have fled to Shanghai, those of the second rank to Hankou, those of the third to Changsha, and those of the fourth to the county towns, while the fifth rank and the still lesser fry surrender to the peasant associations in the villages.

"Here's ten yuan.[2] Please let me join the peasant association," one of the lesser bad gentry will say.

"Ha! Who wants your filthy money?" is the peasants' reply.

Many middle and small landlords, rich peasants, and even some middle peasants, who were formerly opposed to the peasant associations, are now seeking admission. Visiting various places, I often came across such people who pleaded with me, "Mr. Committeeman from the provincial capital, please be my guarantor!"

Under the Qing dynasty, the household census compiled by the local authorities consisted of a regular register and "the other" register, the former for honest people and the latter for burglars, bandits, and similar undesirables. In some places the peasants now use this to scare those who were formerly against the associations. They say, "Put their names down in the other register!"

Afraid of being entered in the other register, such people try various devices to gain admission into the peasant associations. Their minds are entirely set on this, and they do not feel safe until their names are entered in the peasant association register. More often than not the peasant associations turn them down flat, and so they are always on tenterhooks; with the doors of the association barred to them, they are like tramps without a home or, in rural parlance, "mere trash." In short, what was generally sneered at four months ago as the "peasants' gang" has now become something most honorable. Those

---

[2]*yuan:* one Chinese dollar. Its value fluctuated, but it was always a lot of money for poor farmers.

who formerly prostrated themselves before the gentry now all prostrate themselves before the power of the peasants. Everyone, no matter who, admits that the world has changed since last October.

4. It's Terrible and It's Fine

The peasants' revolt in the countryside disturbed the gentry's sweet dreams. When the news from the countryside reached the cities, the urban gentry were immediately in an uproar. When I first arrived in Changsha, I met all sorts of people and picked up a good deal of gossip. From the middle strata of society upwards to the Guomindang right-wingers,[3] there was not a single person who did not sum it all up in the phrase, "It's terrible!" Even very revolutionary people, influenced by the views of the "It's terrible!" school which dominated the climate in the city, became downhearted when they tried to picture the situation in the countryside in their mind's eye and were unable to deny the word "terrible." Even very progressive people could only say, "This kind of thing is inevitable in a revolution, but still it's terrible." In short, no one at all could completely reject this word "terrible." But as I have already said, the fact is that the broad peasant masses have risen to fulfill their historical mission, and that the democratic forces in the countryside have risen to overthrow the forces of feudalism in the countryside. This overthrowing of the feudal forces is the real objective of the national revolution. What Mr. Sun Yatsen wanted, but failed, to accomplish in the forty years he devoted to the national revolution, the peasants have accomplished in a few months. The patriarchal-feudal class of local bullies, bad gentry, and lawless landlords has formed the basis of autocratic government for thousands of years, and is the cornerstone of imperialism, warlordism, and corrupt officialdom. [To overthrow them] is a marvelous feat never before achieved, not just in forty but in thousands of years. It is fine. It is not "terrible" at all. It is anything but "terrible." To give credit where credit is due, if we allot ten points to the accomplishments of the democratic revolution, then the achievements of the city dwellers and the military rate only three points, while the remaining seven points should go to the achievements of the peasants in their rural revolution. "It's terrible!" is obviously a theory for combating the rise

[3] *Guomindang right-wingers:* In 1926 and early 1927, the CCP and GMD cooperated in a united front. Mao was a member of both the GMD and the CCP at the time. Here Mao is criticizing conservative members of the GMD who do not support peasant activism.

of the peasants in the interests of the landlords; it is obviously a theory of the landlord class for preserving the old feudal order and obstructing the establishment of the new democratic order; it is obviously a counterrevolutionary theory. No revolutionary comrade should echo this nonsense. If your revolutionary viewpoint is firmly established, and if you go to the villages and have a look around, you will undoubtedly feel a joy you have never known before. Countless thousands of slaves—the peasants—are there overthrowing their cannibalistic enemies. What the peasants are doing is absolutely right; what they are doing is "fine!" "It's fine!" is the theory of the peasants and of other revolutionaries. Every revolutionary comrade should know that the national revolution requires a great change in the countryside. The Revolution of 1911 did not bring about this change, hence its failure. Now a change is taking place, and this is an important factor for the completion of the revolution. Every revolutionary comrade must support this change or he will be a counterrevolutionary.

5. The Question of "Going Too Far"

Then there is another section of people who say, "Although peasant associations are necessary, their actions at present are undeniably going too far." This is the opinion of the middle-of-the-roaders. But what is the actual situation? True, the peasants are in a sense "unruly" in the countryside. Supreme in authority, the peasant association allows the landlord no say and sweeps away the landlord's prestige. This amounts to striking the landlord down into the dust and trampling on him there. They coined the phrase: "If he has land, he must be a bully, and all gentry are evil." In some of the places even those who own 50 *mu*[4] of fields are called local bullies, and those who wear long gowns are called bad gentry. The peasants threaten, "We will put you in the other register!" They fine the local bullies and bad gentry, they demand contributions from them, and they smash their sedan-chairs.[5] In the case of local bullies and bad gentry who are against the peasant association, a mass of people swarm into their houses, slaughtering their pigs and consuming their grain. They may even loll on the ivory-inlaid beds belonging to the young ladies in the households of the local bullies and bad gentry. At the slightest provocation they make arrests, crown the arrested with tall paper hats, and parade

[4]mu: about one-sixth of an acre
[5]*sedan-chairs:* chairs in which rich people traveled in rural China. Two or four porters would lift and carry the chair, which had sides and curtains.

them through the villages, saying, "You dirty landlords, now you know who we are!" Doing whatever they like and turning everything upside down, they have even created a kind of terror in the countryside. This is what ordinary people call "going too far," or "going beyond the proper limits in righting a wrong," or "really too much." Such talk may seem plausible, but in fact it is wrong. First, the local bullies, bad gentry, and lawless landlords have themselves driven the peasants to this. For ages they have used their power to tyrannize over the peasants and trample them underfoot; that is why the peasants have reacted so strongly. The most violent revolts and the most serious disorders have invariably occurred in places where the local bullies, bad gentry, and lawless landlords perpetrated the worst outrages. The peasants are clear-sighted. Who is bad and who is not, who is the worst and who is not quite so vicious, who deserves severe punishment and who deserves to be let off lightly—the peasants keep clear accounts, and very seldom has the punishment exceeded the crime. Therefore, Mr. Tang Mengxiao[6] also said "The peasants are arresting local bullies and bad gentry, nine of ten arrested deserve it." Secondly, a revolution is not like inviting people to dinner, or writing an essay, or painting a picture, or doing embroidery; it cannot be so refined, so leisurely and gentle, so "benign, upright, courteous, temperate and complaisant."* A revolution is an uprising, an act of violence whereby one class overthrows the power of another. A rural revolution is a revolution in which the peasantry overthrows the power of the feudal landlord class. If the peasants do not use extremely great force, they cannot possibly overthrow the deeply rooted power of the landlords, which has lasted for thousands of years. The rural areas must experience a great, fervent revolutionary upsurge, which alone can rouse the peasant masses in their thousands and tens of thousands to form this great force. All the excessive actions mentioned above [result from] the power of the peasants, mobilized by the great, fervent revolutionary upsurge in the countryside. It was highly necessary for such things to be done in the second period of the peasant movement, the period of revolutionary action. Such actions were extremely necessary during the second period of the peasant movement (the period of revolution).

[6] *Tang Mengxiao:* Tang Shengzhi, a military commander for Hunan's governor Zhao Hengti. In March 1926, Tang succeeded Zhao as governor and worked with the GMD. In early 1927, Tang supported Mao's work, but he later turned against the Communists.
   *These are the qualities that enabled Confucius, according to his disciple Zigong, to obtain information about the government of the countries he visited. See the *Analects*, I, X, 2.

In this period, it was necessary to establish the absolute dominance of the peasants. It was necessary to forbid criticism of the peasant associations. It was necessary to overthrow completely the authority of the gentry, to knock them down and even stamp them underfoot. All excessive actions had revolutionary significance during the second period. To put it bluntly, it is necessary to bring about a brief reign of terror in every rural area; otherwise we could never suppress the activities of the counterrevolutionaries in the countryside or overthrow the authority of the gentry. To right a wrong it is necessary to exceed the proper limits; the wrong cannot be righted without doing so. The argument of this group seems on the surface to differ from that of the group discussed earlier, but essentially they proceed from the same standpoint and likewise voice a landlord theory that upholds the interests of the privileged classes. Since this theory impedes the rise of the peasant movement and so disrupts the revolution, we must firmly oppose it.

II. THE REVOLUTIONARY VANGUARD

1. The Movement of the Riffraff

The right wing of the Guomindang says, "The peasant movement is a movement of the riffraff, a movement of the lazy peasants." This argument has gained much currency in Changsha. When I was in the countryside, I heard the gentry say, "It is all right to set up peasant associations, but the people now running them are no good. They ought to be replaced!" This argument comes to the same thing as what the right-wingers are saying. Both admit that it is all right to have a peasant movement (since the peasant movement has already come into being, no one dare say otherwise), but they regard the people running it as no good. Their hatred is directed particularly against those in charge of the associations at the lower levels, whom they call "riffraff." Those people in the countryside who used to go around in worn-out leather shoes, carry broken umbrellas, wear green gowns, and gamble—in short, all those who were formerly despised and kicked into the gutter by the gentry, who had no social standing, and who were completely deprived of the right to speak, have now dared to lift their heads. Not only have they raised their heads, they have also taken power into their hands. They are now running the township peasant associations (the lowest level of peasant associations), and have turned them into a formidable force. They raise their rough, blackened hands and lay them on the heads of the gentry. They tether the bad gentry with ropes, crown them with tall paper hats, and parade them through the villages. (In Xiangtan and Xiangxiang they call this "parading through the township" and in Liling "parading through the fields.") Every day the coarse, harsh sounds of their denunciations pierce the ears of these gentry. They are giving orders and running everything. They, who used to rank below everyone else, now rank above everybody else—that is what people mean by "turning things upside down."

2. Vanguard of the Revolution or Outstanding Contributors to the Revolution

When there are two different ways of looking at a certain thing, or a certain kind of people, two opposite assessments emerge. "It's terrible!" and "It's fine!" are one example and "riffraff" and "vanguard of the revolution" are another. We said above that the peasants had accomplished a revolutionary task for many years left unaccomplished and had done the principal work in the national revolution. But has this great revolutionary task, this principal work in the revolution, been performed by all the peasants? No. There are three kinds of peasants: the rich, the middle, and the poor peasants. These three categories live in different circumstances and so have different ideas about the revolution. In the first period, what appealed to the rich peasants (those who have surplus money and grain are called rich peasants) was the talk about the Northern Expeditionary Army's sustaining a crushing defeat in Jiangxi, about Chiang Kaishek's being wounded in the leg and flying back to Guangdong, and about Wu Peifu's recapturing Yuezhou.[7] The peasant associations would certainly not last and the Three People's Principles[8] could never prevail, because they had never been heard of before. Thus an official of the township peasant association (generally one of the riffraff type) would walk into the house of a rich peasant, register in hand, and say, "Will you please join the peasant association?" How did the rich peasants reply? "Peasant association? I have lived here for decades, tilling my land. I never saw such a thing before, yet I've managed to live all

[7]These are all rumored setbacks in the Northern Expeditionary Army's military campaigns. Wu Peifu was a northern militarist who at one point reportedly retook the central Chinese city of Yuezhou from the Northern Expeditionary Army.

[8]*Three People's Principles:* socialism, nationalism, and democracy—the goals of Sun Yat-sen and the official ideology of the Northern Expeditionary Army

right," says a rich peasant with a tolerably decent attitude. "I advise you to give it up!" A really vicious rich peasant says, "Peasant association! Nonsense! Association for getting your head chopped off! Don't get people into trouble!" Yet, surprisingly enough, the peasant associations have now been established for several months and have even dared to stand up to the gentry. The gentry of the neighborhood who refused to hand over their opium pipes were arrested by the associations and paraded through the villages. In the county towns, moreover, some big landlords were put to death (such as Yan Rongqiu of Xiangtan and Yang Zhize of Ningxiang). On the anniversary of the October Revolution, at the time of the anti-British rally and of the great celebrations of the victory of the Northern Expedition, tens of thousands of peasants, holding high their banners, big and small, along with their carrying poles and hoes, demonstrated in massive, streaming columns. The rich peasants began to get perplexed and alarmed in their hearts. During the great victory celebrations of the Northern Expedition, they learned that [the city of] Jiujiang had also been taken, that Chiang Kaishek had not been wounded in the leg, and that Wu Peifu had been defeated after all. What is more, they saw "Long live the Three People's Principles!" "Long live the peasant associations!" "Long live the peasants!" and so on and so forth clearly written on the red and green proclamations (slogans).

"What?" wondered the rich peasants, greatly perplexed and alarmed, "'Long live the peasants!' Are these people now to be regarded as emperors?"* So the peasant associations are putting on grand airs. People from the associations say to the rich peasants:

"We'll enter you in the other register!"

"In another month, the admission fee will be ten yuan a head!"

Only under the impact and intimidation of all this are the rich peasants tardily joining the associations, some paying fifty cents or one yuan for admission (the regular fee being a mere ten coppers), some securing admission only after asking other people to put in a good word for them. But there are quite a number of diehards who have not joined to this day. When the rich peasants join the associations, they generally enter the name of some sixty- or seventy-year-old member of the family, for they are in constant dread of conscription. After joining, the rich peasants are not keen on doing any work for the association. They remain inactive throughout. How about the middle peasants?

*The most common expression in Chinese for "Long live!" is *wansui*, literally "ten thousand years," which was used of the emperor.

(Those who do not have any surplus money and rice, are not in debt, and are able to assure themselves of clothing, food, and shelter every year are called the middle peasants.) The attitude of the middle peasants is a vacillating one. They think that the revolution will not bring much good to them. They have rice cooking in their pots and no creditors knocking on their doors at midnight. They, too, judging a thing by whether it ever existed before, knit their brows and think to themselves, "Can the peasant association really last?" "Can the Three People's Principles prevail?" Their conclusion is, "Afraid not!" They imagine it all depends on the will of Heaven and think, "A peasant association? Who knows if Heaven wills it or not?" In the first period, people from the association would call on a middle peasant, register in hand, and say, "Will you please join the peasant association?" The middle peasant replied, "There's no hurry!" It was not until the second period, when the peasant associations were already exercising great power, that the middle peasants came in. Even though they are somewhat better in the peasant associations than the rich peasants, they are never very enthusiastic, and retain their vacillating attitude. The only kind of people in the countryside who have always put up the bitterest fight are the poor peasants. From the period of underground work straight through to the period of open activity, it is they who have fought. As for organization, it is they who are organizing things there, and as for revolution, it is they who are making revolution there. They alone are the deadly enemies of the local tyrants and evil gentry, and they strike them without the slightest hesitation. They alone are capable of carrying out the work of destruction. They say to the rich and middle peasants:

"We joined the peasant association long ago, why are you still hesitating?"
The rich and the middle peasants answer mockingly:
"What is there to keep you from joining? You people have neither a tile over your heads nor a speck of land under your feet!"

It is true that the poor peasants are not afraid of losing anything. They are the disinherited or semidisinherited in rural life. Some of them really have "neither a tile over their heads nor a speck of land under their feet." What, indeed, is there to keep them from joining the associations? According to the survey of Changsha county, the poor peasants comprise 70 percent, the middle peasants 20 percent, and the rich peasants 10 percent. The 70 percent, the poor peasants, may be subdivided into two categories, the utterly destitute and the less destitute. The "utterly destitute" are the completely dispossessed, that is, people who have neither land nor capital, are without any means of livelihood, and are forced to leave home and become mercenaries or hired laborers and wandering beggars, or commit crimes and become robbers and thieves. They make up 20 out of the 70 [percent]. The less destitute are the partially dispossessed, that is, people with just a little land or a little capital who eat up more than they earn and live in toil and distress the year round, such as the handicraftsmen, the tenant-peasants (not including the rich tenant-peasants), and the semitenant-peasants. These make up 50 out of the 70 [percent]. (The number of poor peasants in other counties may be smaller than in Changsha, but there should not be a big discrepancy.) This great mass of poor peasants constitute the backbone of the peasant associations, the vanguard in overthrowing the feudal forces, and the foremost heroes who have performed the great revolutionary task which for long years was left undone. Without the poor peasant class (the riffraff, as the gentry call them), it would never have been possible to bring about the present revolutionary situation in the countryside, or to overthrow the local bullies and bad gentry and to complete the democratic revolution. The poor peasants (especially the portion who are utterly destitute), being the most revolutionary group, have gained the leadership of the peasant associations. In both the first and second periods almost all the chairmen and committee members in the peasant associations at the lowest level (i.e., the township associations) were poor peasants (of the officials in the township associations in Hengshan county the class of the utterly destitute comprise 50 percent, the class of the less destitute 40 percent, and poverty-stricken intellectuals 10 percent). This leadership by the poor peasants is extremely necessary. Without the poor peasants there would be no revolution. To deny their role is to deny the revolution. To attack them is to attack the revolution. From beginning to end, the general direction they have given to the revolution has never been wrong. They have discredited the local bullies and bad gentry. They have knocked down the local bullies and bad gentry, big and small, and trampled them underfoot. Many of their "excessive" deeds in the period of revolutionary action were in fact the very things the revolution required. Some county governments, county headquarters of the Guomindang, and county peasant associations in Hunan have already made a number of mistakes; some have even sent soldiers to arrest officials of the lower-level associations at the landlords' request. A good many chairmen and committee members of township associations in Hengshan and Xiangxiang counties have been thrown in jail. This mistake is very serious and unintentionally feeds the arrogance of the reactionaries. To judge whether or not it is a mistake, you have only to see how joyful the lawless landlords become and how reactionary sentiments grow whenever the chairmen or committee members of local peasant associations are arrested. We must combat the counterrevolutionary slogan of a "movement of riffraff" and a "movement of lazy peasants," but at the same time we should be especially careful not to help the local bullies and bad gentry (however unintentionally) in their attacks on the leadership of the poor peasant class. In fact, though a few of the poor peasant leaders undoubtedly did "gamble, play cards, and not earn their living by hard work," most of them have changed by now. They themselves are energetically prohibiting gambling and suppressing banditry. Where the peasant association is powerful, gambling has stopped altogether and the peril of banditry has vanished. In some places it is literally true that people do not take articles left by the wayside and that doors are not bolted at night. According to the Hengshan survey, 85 percent of the poor peasant leaders have made great progress and have proved themselves capable and hard-working. Only 15 percent retain some bad habits. The most one can call them is "an unhealthy minority," and we must not echo the local bullies and bad gentry in undiscriminatingly condemning them as "riffraff." As to dealing with the "unhealthy minority," we can proceed only under the peasant associations' own slogan of "strengthen discipline," by conducting propaganda among their masses, by training the "unhealthy minority," and by improving the discipline of the associations; in no circumstances should soldiers be arbitrarily sent to make such arrests as would weaken the faith [in] the poor peasants and feed the arrogance of the local bullies and bad gentry. This point requires careful attention.

III. PEASANTS AND THE PEASANT ASSOCIATIONS

Most critics of the peasant associations allege that they have done a great many bad things. I have already pointed out in the preceding two sections that the peasants' attack on the local bullies and bad gentry is entirely revolutionary behavior and in no way blameworthy. But the peasants have done a great many things, and we must closely examine all their activities to see whether or not what they have done is really all bad, as is being said from without. I have summed up their activities of the last few months; in all, the peasants under the command of the peasant associations have the following fourteen great achievements to their credit.

1. Organizing the Peasants under Peasant Associations

This is the first great thing the peasants have achieved. In counties such as Xiangtan, Xiangxiang, and Hengshan, nearly all the peasants are organized and there is hardly a remote corner where they are not on the move; these are the best places. In some counties like Yiyang and Huarong, the bulk of the peasants have arisen, with only a small section not yet arisen; these places are in the second grade. In other counties, like Chengbu and Lingling, while a small section has arisen, the bulk of the peasants have still not arisen; these places are in the third grade. Western Hunan, which is under the control of Yuan Zuming, has not yet been reached by the associations' propaganda, and the peasants of many of its counties have completely failed to rise; these form a fourth grade. Roughly speaking, the counties in central Hunan, with Changsha as the center, are the most advanced, those in southern Hunan come second, and western Hunan is only just beginning to organize. According to the figures compiled by the provincial peasants' association last November, organizations with a total membership of 1,367,727 have been set up in thirty-seven of the province's seventy-five counties. Of these members, about one million were organized during the time of October and November when the power of the associations rose high, while up to September the membership had only been 300,000 to 400,000. Then came the two months of December and January, and the peasant movement continued its brisk growth. By the end of the month the membership must have reached at least two million. As a family generally enters only one name when joining and has an average of five members, the mass following must have reached ten million. This astonishing and accelerating rate of expansion explains why the local bullies, bad gentry, and corrupt officials have been isolated; why society has been amazed at how different the world was before and after; and why a great revolution has been wrought in the countryside. This is the first great thing that the peasants have achieved under the command of the peasant associations.

The table [on pp. 56–58] gives the membership of the peasant associations in all the counties in Hunan province as of last November [1926].

2. Dealing Political Blows to the Landlords

After the peasants are organized, the first thing they do is to smash the political prestige of the landlord class, and especially of the local bullies and bad gentry, that is, to pull down the power and influence of the landlords and build up the power and influence of the peasants in rural society. This is a most serious and urgent struggle; it is the central struggle in the second period, the period of revolution. If this struggle is not victorious, there can be no possibility of victory in any of the economic struggles, such as the struggle for rent and interest reduction, or for capital and land, and so on. In many places in Hunan like Xiangxiang, Hengshan, and Xiangtan counties, this is of course no problem since the power of the landlords has been overturned and the peasants constitute the sole power. But in counties like Liling, there are still some places (such as the two western and southern counties of Liling) where the power of the landlords seems weaker than that of the peasants but, because the political struggle has not been sharp, landlord power is in fact surreptitiously opposing peasant power. In such places it is still too early to say that the peasants have gained political victory; they must wage the political struggle more vigorously until the power of the landlords is completely cast down. . . .

[In the remaining part of this item, Mao catalogs the methods used by peasants to attack landlord power in Hunan: auditing accounts, imposing fines, levying contributions, holding minor and major demonstrations, parading landlords in dunce caps, jailing, banishment, and execution. In the next item, Item 3, Mao covers the economic powers of the peasant associations, including keeping grain in the county, limiting rents, protecting tenants' rights, and lowering interest on loans.]

4. Overthrowing the Feudal Politics of the Local Bullies and Bad Gentry in the Rural Areas—Smashing the Districts and the Townships

The old organs of rural administration in the districts and townships, and especially at the district level (namely just below the county level), used to be almost exclusively in the hands of the local bullies and bad gentry. They had jurisdiction over a population of from ten to fifty or sixty thousand people. They had their own independent armed forces, such as the township defense corps; their own independent fiscal

Comparative Table of Peasant Association Membership by County

NAME OF COUNTY	NO. OF QU ASSNS[a]	NO. OF XIAN ASSNS[b]	AGR. LABORERS	SHARE CROPPERS	SEMI-OWNER PEASANTS	OWNER PEASANTS	HANDI-CRAFT WORKERS	PRIMARY SCHOOL TEACHERS	SMALL MERCHANT	WOMEN	OTHER	NO. OF MEMBERS
							SOCIAL STATUS OF MEMBERS					
Xiangxiang	44	499	16,400	91,500	41,000	13,100	28,000	450				190,544
Xiangyin		67	15,857	87,590	52,635	14,793	12,514	151	634	57	400	176,000
Liuyang	21	568										139,190
Xiangtan	17	450	27,000	54,100	12,400	8,460	7,400	1,100				120,460
Hengyang	23	244	17,358	37,725	7,532	5,628	6,135	2,256			1,579	88,223
Changsha	12	640	17,527	25,948	9,131	5,381	4,915	1,425	1,463	643		66,415
Anhua	15	120										62,300
Liling	15	323	6,746	35,460	6,920	3,998	3,643	230	601	195	683	58,476
Ningxiang	18	400	5,000	20,000	10,000	4,000	8,400	600	100	16		58,000
Chenxian	14	696	19,725	26,898	2,124	2,550	5,711	118				57,262
Hengshan	13	203	3,623	16,933	2,965	2,174	3,328	332			1,133	30,016
Suburbs		169	9,509	10,646	3,563	2,893	1,794	254	582	156		29,475
Linwu	6	32	2,183	10,143	4,146	2,291	933					20,000
Youxian		29										18,400
Yiyang	7	67	1,568	5,017	6,586	1,586	784	32	126			15,680
Huarong	6	49	2,000	6,595	2,453	1,887	401	1,216				14,652
Yizhang	10	185	1,438	8,936	1,637	1,283	802	87				14,183
Laiyang	9	149	1,145	6,865	2,684	1,844	342	66				12,946
Linli	6	49	2,000	3,000	2,400	4,000	200	60				11,660
Chaling	4	124	500	7,000	2,500	1,000	200	60				11,260
Yongxing	16	107	1,200	2,800	4,020	2,200	200	30				10,450

56

County												Total
Pingjiang	17	162	1,023	4,298	1,781	1,612	1,093	214	85	4	42	10,152
Xinning	9	25	1,722	6,533	858	375	184	74				9,746
Changde	3	59	890	2,800	2,080	3,500	310	65				9,545
Baoqing	7	136	1,438	2,367	1,481	1,744	771					9,377
Wugang	8	40	1,800	4,500	900	900	900	900				9,000
Rucheng	6	46	406	4,195	2,957	1,228	41	38				8,865
Hanshou	69	1,125	3,276	1,047	1,378	228	33	61	78			7,226
Nanxian	6	49	1,384	4,064	907	406	69	89	45		36	7,000
Zhuping Lu		21	997	3,152	732	539	687	50	297	10		6,464
Xinhua	6		1,526	3,246	497	424	472	202				6,377
Guiyang	4	52	445	274	1,673	1,525	402	24				6,245
Qiyang	15	70										6,000
Lingxian	12	48	1,312	1,917	601	492	546	218	382			5,468
Zixing	5	79	2,148	1,123	891	341	699	122				5,324
Guidong	7	95	816	1,156	1,022	1,507	94	62	204	297	25	5,193
Xintian	8	47	456	2,927	955	488	299	25				5,150
Changning		78	486	2,378	823	536	96	12		34		4,549
Cili	11	48	263	1,550	601	1,806	236	40	105		178	4,496
Linxiang	7	95	624	995	847	1,195	47	31		152	81	4,077
Taoyuan	7	36										4,000
Yuanjiang	3	19	241	1,174	520	1,615	243	46				3,839
Lanshan	4	51	765	1,499	604	385	41	21			35	3,350
Lixian	4	16	597	1,033	389	249	215	66				2,549
Jiahe	3	27	295	598	588	850	89	32				2,452

[a] A *qu* is a subdistrict of a county.
[b] A *xiang* is a subdistrict of a *qu*.

Comparative Table of Peasant Association Membership by County (continued)

NAME OF COUNTY	NO. OF QU ASSNS[a]	NO. OF XIAN ASSNS[b]	SOCIAL STATUS OF MEMBERS									NO. OF MEMBERS
			AGR. LABORERS	SHARE CROPPERS	SEMI-OWNER PEASANTS	OWNER PEASANTS	HANDI-CRAFT WORKERS	PRIMARY SCHOOL TEACHERS	SMALL MERCHANT	WOMEN	OTHER	
Anxiang	6	13	280	760	680	440	120	18				2,298
Yongming	5	31	58	522	1,150	420	21	11				2,182
Yueyang	7	47	136	830	410	558	65	11				2,010
Xupu	2	11	540	775	331	204	108	7				1,965
Daoxian	13	39	136	540	282	403	56	18				1,435
Luxi	3	17	102	350	520	240	82	12				1,306
Suining	4	15	121	314	332	297	13	34				1,111
Ningyuan	8	13	86	480	105	159	42	20				892
Chengbu	1	8	130	195	372	101	74	13	14			889
Lingling	4	15	23	133	167	251	8	28	29		58	697
Mayang		9	130	348	36	19	4	3	21	21	48	630
Zhijiang		4		118	76	73		7				274
Total	461	6,867										1,367,727

[a] A *qu* is a subdistrict of a county.
[b] A *xiang* is a subdistrict of a *qu*.

58

powers, such as the power to levy taxes; and their own judiciary, which could freely arrest, imprison, try, and punish the peasants and so on. The bad gentry who ran these organs of rural administration were virtual monarchs of the countryside. Comparatively speaking, the peasants were not so much concerned with the president of the republic, the provincial military governor, or the county magistrate; their real "bosses" were these rural monarchs. A mere snort, and the peasants all knew they had to watch their step. As a consequence of the present revolt in the countryside, the authority of the landlord class has been struck down everywhere, and the organs of rural administration dominated by the local bullies and bad gentry have naturally collapsed in its wake. The heads of the districts and the townships all steer clear of the people, dare not show their faces and hand all local matters over to the peasant associations. They put people off with the remark, "It is none of my business!"

Whenever their conversation turns to the heads of the districts and the townships, the peasants say angrily, "That bunch! They are finished!"

Yes, the term "finished" truly describes the state of the old organs of rural administration wherever the storm of revolution has raged.

5. Overthrowing the Armed Forces of the Landlords and Establishing Those of the Peasants

The armed forces of the landlord class were smaller in central Hunan than in the western and southern parts of the province. An average of 600 rifles for each county would make a total of 45,000 rifles for all the seventy-five counties; there may, in fact, be more than this number. In the southern and central parts where the peasant movement is being developed, the landlord class cannot hold its own because of the overwhelming momentum with which the peasants have risen, and its armed forces have largely capitulated to the peasant associations and taken the side of the peasants; examples of this are to be found in such counties as Ningxiang, Pengjiang, Liuyang, Changsha, Liling, Xiangtan, Xiangxiang, Anhua, Hengshan, and Hengyang. In some counties such as Baoqing and so on, a small number of the landlords' armed forces are taking a neutral stand, though still with a tendency to capitulate. Another small section are opposing the peasant associations, but the peasants are attacking them and may wipe them out before long, as, for example, in such counties as Yichang, Linwu, and Jiahe. At the present time, stronger measures are being taken against these forces, which may all be eradicated soon. The armed forces thus taken over from the reactionary landlords are all being reorganized into a "standing household militia" and are under the new organs of rural self-government, which are organs of the political power of the peasantry. This "taking over these old armed forces" is one part of building up an armed force of the peasantry. Even though some of them are still struggling, the various counties in southern and central Hunan have no problems anymore. There are some problems only in western Hunan. In addition, there is a new way for establishing an armed force of the peasants, which is through the setting up of spear corps under the peasant associations. The spears have pointed, double-edged blades mounted on long shafts, and there are now 100,000 of these weapons in Xiangxiang county alone. Other counties such as Xiangtan, Hengshan, Liling, and Changsha have 70,000 to 80,000, or 50,000 to 60,000, or 30,000 to 40,000 each. In every county where there is a peasant movement, the spears are spreading rapidly. These peasants thus armed form an "irregular household militia." This multitude equipped with spears, which is larger than the old armed forces mentioned above, is a newborn "thing," the mere sight of which makes the local tyrants and evil gentry shiver. The revolutionary authorities in Hunan should see to it that this kind of thing is built up on a really extensive scale among the more than 20 million peasants in the seventy-five counties of the province, that every peasant, whether young or in his prime, possesses a spear, and that no restrictions are imposed as though a spear were something dreadful. Anyone who is scared at the sight of the spear corps is indeed a weakling! Only the local bullies and bad gentry are frightened of them, but no revolutionaries should take fright.

6. Overthrowing the Political Power of the County Magistrate and His Bailiffs

That only if the peasants rise can the county government be cleaned up has already been proved in Haifeng, Guangdong province. On this occasion in Hunan, we have obtained further ample proof. In a county that is under the sway of the local bullies and bad gentry, the magistrate, whoever he may be, is always a corrupt official. In a county where the peasants have risen there is clean government, whoever the magistrate may be. In the counties I visited, the magistrates had to consult the peasant associations on everything in advance. In counties where the power of the peasant movement was very strong, the word of the peasant association worked miracles. If the peasant association demanded the arrest of a local bully in the morning, the magistrate dared not delay till noon; if they demanded it by noon, he dared not delay till the afternoon. When the power of the peasants was just beginning to make itself felt in the countryside, the magistrate worked in league with the local bullies and bad gentry. When the peasants' power grew till it matched that of the landlords, the magistrate took the position of trying to accommodate both sides, accepting some of the peasant association's suggestions while rejecting others. The remark that "the word of the peasants works miracles" applies only when the power of the landlords has been completely beaten down by that of the peasants. At present the political situation in counties such as Xiangxiang, Xiangtan, Liling, and Hengshan is as follows:

a. All decisions are made by a joint council consisting of the magistrate and the representatives of the revolutionary mass organizations. The council is convened by the magistrate and meets in his office. In some counties it is called the "joint council of public bodies and the local government," and in others the "council of county affairs." Besides the magistrate himself, those attending but not voting are the representatives of the county peasant association, trade union council, merchant association, women's association, school staff association, student association, and Guomindang party office. At such council meetings the magistrate is influenced by the views of the public organizations and "invariably does their bidding." The adoption of a democratic committee system of county government does not, therefore, present the slightest problem in Hunan. The present county governments are already quite "democratic" both in form and substance. This situation has been brought about only in the last two or three months, that is, since the peasants have risen all over the countryside and overthrown the power of the local bullies and bad gentry. It has now come about that the magistrates, seeing their old props collapse and needing new props to retain their posts, have begun to curry favor with the public organizations, and the situation has changed as described above.

b. The judicial assistant has scarcely any cases to handle. The judicial system in Hunan remains one in which the county magistrate is concurrently in charge of judicial affairs, with an assistant to help him in handling cases. To get rich, the magistrate and his underlings used to rely entirely on "collecting taxes and levies, procuring men and provisions for the armed forces," and "extorting money in civil and criminal lawsuits by confounding right and wrong," the last being the most regular and reliable source of income. In the last few months, with the downfall of the local bullies and bad gentry, all the legal pettifoggers have disappeared. What is more, the peasants' problems, big and small, are now all settled in the peasant associations at the various levels. Thus the county judicial assistant simply has nothing to do. The one in Xiangxiang told me, "When there were no peasant associations, an average of sixty civil or criminal suits were brought to the county government each day; now it receives an average of only four or five suits a day." So it is that the purses of the magistrates and their underlings perforce remain empty.

c. The armed guards, the police, and the bailiffs all keep out of the way and dare not go near the villages to practice their extortions. In the past the people in the villages were afraid of the people in the towns, but now the people in the towns are afraid of the people in the villages. In particular the vicious curs kept by the county government—the police, the armed guards, and the bailiffs—are afraid of going to the villages, or if they do so, they no longer dare to practice their extortions. They tremble at the sight of the peasants' spears.

7. Overthrowing the Clan Authority of the Ancestral Temples and Clan Elders, the Religious Authority of Town and Village Gods, and the Masculine Authority of Husbands

A man in China is usually subjected to the domination of three systems of authorities: (1) the state system (political authority), ranging from the national, provincial, and county government down to that of the township; (2) the clan system (clan authority), ranging from the central ancestral temple and its branch temples down to the head of the household; and (3) the supernatural system (religious authority), ranging from the King of Hell down to the town and village gods belonging to the nether world, and from the Emperor of Heaven down to all the various gods and spirits belonging to the celestial world. As for women, in addition to being dominated by these three, they are also dominated by men (the authority of the husband). These four authorities—political, clan, religious, and masculine—are the embodiment of the whole feudal-patriarchal ideological system, and are the four thick ropes binding the Chinese people, particularly the peasants. How the peasants have overthrown the political authority of the landlords in the countryside has been described above. The political authority of the landlords is the backbone of all the other systems of authority. With the politics of the landlords overturned, the clan authority, the religious authority, and the authority of the husband all begin to totter. Where the peasant association is powerful, the clan elders and administrators of temple funds no longer dare oppress those lower in the clan hierarchy or embezzle clan funds. The worst clan elders and administrators, being local bullies, have been thrown out. No one any longer dares to practice the corporal and capital punishments that used to be inflicted in the ancestral temples, such as flogging, drowning, and burying alive. The old rule barring women and poor people from the banquets in the ancestral temples has also been broken. The women of Baiguo in Hengshan county gathered in force and swarmed into their ancestral temple, firmly planted their backsides on the seats, and joined in the eating and drinking, while the venerable clan bigwigs had willy-nilly to let them do as they pleased. At another place, where poor peasants had been excluded from temple banquets, a group of them flocked in and ate and drank their fill, while the local bullies and bad gentry and other long-gowned gentlemen all took to their heels in fright. Everywhere religious authority totters as the peasant movement develops. In many places the peasant associations have taken over the temples of the gods as their offices. Everywhere they advocate the appropriation of temple property for peasant schools and to defray the expenses of the associations, calling it "public revenue from superstition." In Liling county, prohibiting superstitious practices and smashing idols have become quite the vogue. In its northern districts the peasants have prohibited the incense-burning processions to propitiate the god of pestilence. There were many idols in the Daoist temple on Fubo Hill in Lukou, but when extra premises were needed for the district party offices [of the Guomindang], they were all piled up in a corner, big and small together, and no peasant raised any objection. Since then, sacrifices to the gods, the performance of religious rites, and the offering of sacred lamps have rarely been practiced when a death occurs in a family. Because the initiative in this matter was taken by the chairman of the peasant association, Sun Xiaoshan, he is hated by the local Daoist priests. In the Longfeng Nunnery in the North Third District, the peasants and primary school teachers chopped up the wooden idols and actually used the wood to cook meat. More than thirty idols in the Dongfu Monastery in the Southern District were burned by the students and peasants together, and only two small images of Bao Gong were snatched up by an old peasant who said, "Don't commit a sin!" Everywhere it has always been the case that only the older peasants and the women believe in the gods; all the younger peasants do not. Since the latter control the associations, the overthrow of religious authority and the eradication of superstition are going on everywhere. As to the authority of the husband, this has always been weaker among the poor peasants because, out of economic necessity, their womenfolk have to do more manual labor than the women of the richer classes and therefore have more say and greater power of decision in family matters. In sexual matters, they also have relatively more freedom. Among the poor peasants in the countryside, triangular and multilateral relationships are almost universal. With the increasing bankruptcy of the rural economy in recent years, the basis for men's domination over women has already been weakened. With the rise of the peasant movement, the women in many places have now begun to organize rural women's associations; the opportunity has come for them to lift up their heads, and the authority of the husband is getting shakier every day. In a word, the whole feudal-patriarchal ideological system is tottering with the growth of the peasants' power. But in the past and at the present time, the peasants are concentrating entirely on destroying the landlords' political authority. Wherever it has been wholly destroyed, they are beginning to press their attack in the three other spheres of the clan, the gods, and male domination. But such attacks have only just "begun," and there can be no thorough overthrow of all three until the peasants have won complete victory in the economic fight. Therefore, our present task is to lead the peasants to put their greatest efforts into the political struggle, so that the landlords' authority is entirely overthrown. The economic struggle should follow immediately, so that the economic problems of the poor peasants may be fundamentally solved. As for smashing the clan system, superstitious ideas, and one-sided concepts of chastity, this will follow as a natural consequence of victory in the political and economic struggles. If too much of an effort is made arbitrarily and prematurely to abolish these things, then the local bullies and bad gentry will seize the pretext to put forward such slogans as "the peasant association has no piety towards ancestors," "the peasant association is blasphemous and is destroying religion," and "the peasant association stands for the communization of wives," all for the purpose of undermining the peasant movement. A case in point is the recent events at Xiangxiang, Hunan, and Yangxin, Hubei, where the landlords exploited the opposition of some peasants to smashing idols. It is the peasants who made the idols, and when the time comes they will cast the idols aside with their own hands; there is no need for anyone else to do it for them prematurely. Our propaganda policy in such matters is, "Draw the bow but do not release the arrow, having seemed to leap."* The idols should be removed by the peasants themselves, the ancestral tablets should be smashed by the peasants themselves, the temples to martyred virgins and arches for chaste and filial widows and daughters-in-law should be demolished by the peasants themselves.

While I was in the countryside, I did some propaganda against superstition among the peasants. I said: "If you believe in the Eight Characters,[9] you hope for good luck; if you believe in geomancy,[10] you hope to benefit from the location of your ancestral graves. This year within the space of a few months the local bullies, bad gentry, and corrupt officials have all fallen from power. Is it possible that until a few months ago they all had good luck and enjoyed the benefit of well-sited ancestral graves, while suddenly in the last few months their luck has turned and their ancestral graves have ceased to exert a beneficial influence?

"The local bullies and bad gentry jeer at your peasant association and say, 'How odd! Today, the world is a world of committeemen. Look, you can't even go to pass water without bumping into a committeeman!' Quite true, the towns and the villages, the peasant associations and the labor unions, the Guomindang and the Communist Party, all without exception have their executive committee members—it is indeed a world of committeemen. But is this caused by the Eight Characters and the location of the ancestral graves? How strange! The Eight Characters of all the poor wretches in the countryside have suddenly turned auspicious! And their ancestral graves have suddenly started exerting beneficial influences!

"The gods? Worship them by all means. But if you had only Lord Guan and the Goddess of Mercy and no peasant association, could you have overthrown the local tyrants and evil gentry? The 'gods' and 'goddesses' are indeed miserable objects. You have worshipped them for several thousand years, and they have not overthrown a single one of the local bullies or bad gentry for you! Now you want to have your

---

*Mao here takes his text from the *Mencius,* VII, I, XLI, 3. The moral that Mao wished to draw from the passage is clear, in any case: the master illustrates the action to be taken, driving home the message with dramatic gestures, but leaves it to the disciples to carry out the action.

[9]*Eight Characters:* a written record of the astrological moment of a person's birth and thus his or her fate

[10]*geomancy:* a practice in which the placement of household furniture, buildings, and even family graves is chosen to produce auspicious effects; also known as feng shui

rent reduced. Let me ask you, what method will you use? Will you place your faith in the gods, or in the peasant associations?"

When I spoke these words, the peasants laughed, and in the midst of their laughter, I imagined that the gods and idols all fled from sight.

8. Spreading Political Propaganda

Even if ten thousand schools of law and political science had been opened, could they have brought as much political education to the people, men and women, young and old, all the way into the poorest and remotest corners of the countryside, as the peasant associations have done in so short a time? I think they certainly could not have. Down with imperialism! Down with the warlords! Down with the corrupt officials! Down with the local bullies and bad gentry!—these political slogans have grown wings, they have found their way to the young, the middle-aged, and the old, to the women and children in countless villages, they have penetrated into their minds and flowed back from their minds into their mouths. Suppose, for example, you watch a group of children at play. If one gets angry with another, if he glares, stamps his foot, and shakes his fist, you will then immediately hear from the other the shrill cry: "Down with imperialism!"

In the Xiangtan area, when the children who pasture the cattle get into a fight, one will take the part of Tang Shengzhi and the other that of Ye Kaixin.[11] When, after a while, one is defeated and runs away with the other chasing him, it is the pursuer who is Tang Shengzhi and the pursued Ye Kaixin. As to the song "Down with the Imperialist Powers...," of course almost every child in the towns can sing it, and now many village children can sing it too. Some of the peasants can also recite a little of Mr. Sun Yatsen's Testament. They pick out from it the terms "freedom," "equality," "the Three People's Principles," and "unequal treaties" and apply them, if rather crudely, in their life. When somebody who looks like one of the gentry encounters a peasant on the road and stands on his dignity, refusing to make way along a pathway, the peasant will say angrily, "Hey, you local bully, don't you know the Three People's Principles?"

Formerly, when the peasants from the vegetable farms on the outskirts of Changsha entered the city to sell their produce, they used to be pushed around by the police. Now they can find a weapon, which is no other than the Three People's Principles. When a policeman strikes or swears at a peasant from a vegetable farm, the peasant from the

[11]*Ye Kaixin:* a competitor of Tang Shengzhi in Hunan

vegetable farm immediately answers back by invoking the Three People's Principles and the policeman has not a word to say. Once in Xiangtan, when a district peasant association and a township peasant association could not see eye to eye about a certain matter, the chairman of the township association declared: "Down with the district peasant association's unequal treaties!"

The spread of political propaganda throughout the rural area is entirely an achievement of the peasant associations. Simple slogans, cartoons, and speeches have produced such a widespread and speedy effect among the peasants that it is as though every one of them had been to a political school. According to the reports of comrades engaged in rural work, the influence of extensive political propaganda was to be found in the three great mass movements: the anti-British demonstration, the celebration of the October Revolution, and the victory celebration for the Northern Expedition. In these movements, political propaganda was conducted extensively wherever there were peasant associations, arousing the whole countryside. Consequently, the impact was very great. From now on, care should be taken to make use of every opportunity gradually to enrich the content and clarify the meaning of the simple slogans mentioned above!

9. Peasant Bans and Prohibitions

When the peasant associations establish their authority in the countryside, the peasants begin to forbid strictly or to restrict the things they dislike. Gaming, gambling, and opium smoking are the three things that are most strictly forbidden.

> Gaming: Where the peasant association is powerful, mahjong,[12] dominoes, and card games are wholly banned.
> The peasant association in the Fourteenth District of Xiangxiang burned two basketfuls of mahjong [pieces].
> If you go to the countryside, you will find none of these games played; anyone who violates the ban is promptly and strictly punished.
> Gambling: Former hardened gamblers are now themselves forcefully suppressing gambling; this abuse, too, has been swept away in places where the peasant association is powerful.
> Opium smoking: The prohibition is extremely strict. When the peasant association orders the surrender of opium pipes, no

[12]*mahjong:* a board game using small tiles on which bets can be made

one dares to raise the least objection. In Liling county, one of the bad gentry who did not surrender his pipes was arrested and paraded through the villages.

The peasants' campaign to "disarm the opium smokers" is no less impressive than the disarming of the troops of Wu Peifu and Sun Chuanfang by the Northern Expeditionary Army. Quite a number of venerable fathers of officers in the revolutionary army, old men who were opium addicts and inseparable from their pipes, have been disarmed by the "emperors" (as the peasants are called derisively by the bad gentry). The "emperors" have banned not only the growing and smoking of opium, but also trafficking in it. A great deal of the opium transported from Guizhou to Jiangxi via the various counties of Baoqing, Xiangxiang, Yuoxian, and Liling has been intercepted on the way and burned. This has affected government revenues. As a result, out of consideration for the army's need for funds in the Northern Expedition, the provincial peasant association ordered the associations at the lower levels "temporarily to postpone the ban on opium traffic." This, however, has upset and displeased the peasants.

There are many other things besides these three that the peasants have prohibited or restricted, the following being some examples:

The flower drum. An obscene and vulgar local opera. Its performances are forbidden in many places.

Sedan-chairs. In many counties, especially Xiangxiang, there have been cases of smashing sedan-chairs. A prohibition on taking sedan-chairs has become a vogue. The only people who can take sedan-chairs are the peasant movement officials; otherwise, they will be smashed. The peasants, detesting the people who use this conveyance, are always ready to smash the chairs, but the peasant associations forbid them to do so. Peasant movement officials tell the peasants, "If you smash the chairs, you only save the rich money and lose the carriers their jobs. And the carriers will be out of a job if they have no work to do. Will that not hurt yourselves? Seeing the point, the peasants answer, "That's right." They then adopt a new [policy on] sedan chairs—"to increase considerably the fares charged by the chair-carriers" so as to penalize the rich.

Distilling and sugar-making. The use of grain for distilling spirits and making sugar is everywhere prohibited, and therefore the distillers and sugar refiners are constantly complaining. Distilling is not banned in Futianpu, Hengshan county, but prices are fixed very low, and the wine and spirits dealers, seeing no prospect of profit, have had to stop it.

Pigs. The number of pigs a family can keep is limited, for they consume grain.

Chickens and ducks. In Xiangxiang county the raising of chickens and ducks is prohibited, but the women object. In Hengshan county, each family in Yangtang is allowed to keep only three chickens, and in Futianpu five chickens. In many places the raising of ducks is completely banned, for ducks not only consume grain but also ruin the rice plants and so are worse than chickens.

Feasts. Sumptuous feasts are generally forbidden. In Shaoshan, Xiangtan county, it has been decided that guests are to be served only three kinds of animal food, namely, chicken, fish, and pork. It is also forbidden to serve bamboo shoots, kelp, and lentil noodles. In Hengshan county it has been resolved that eight dishes and no more may be served at a banquet, and not even one more is allowed. Only five dishes are allowed in the East Third District in Liling county, and only three meat and three vegetable dishes in North Second District, while in the West Third District New Year feasts are forbidden entirely. In Xiangxiang county, there is a ban on all "egg-cake feasts," which are by no means sumptuous. When Tie Jiawan in the Second District gave an "egg-cake feast" at a son's wedding, the peasants, seeing the ban violated, swarmed into the house and destroyed the "egg-cake feast." In the town of Jiamuo, Xiangxiang county, the people have refrained from eating expensive foods and use only fruit when offering ancestral sacrifices.

Oxen. Oxen are treasured possessions of the peasants in the South. "Slaughter an ox in this life and you will be an ox in the next" has become almost a religious tenet; oxen must never be killed. Before the peasants had power, they could only appeal to religious taboos in opposing the slaughter of cattle and had no real power to ban it. People in the towns always want to eat beef, and therefore people in the towns always want to kill cattle. Since the rise of the peasant associations, their real jurisdiction has extended even to the cattle, and they have prohibited the slaughter of cattle in the towns. Of the six butcheries that formerly existed in the county town of Xiangtan, five are now closed and the remaining merchant slaughters only enfeebled or disabled animals. The slaughter of cattle is totally prohibited throughout Hengshan county. No one in the county town dares slaughter either. A peasant whose ox fell from a high place, broke a leg, and is now disabled dared not kill it. He consulted the peasant association and got their permission before he dared kill it. When the chamber of commerce of Zhuzhou rashly slaughtered a cow, the peasants one day swarmed into town and demanded an explanation. As a result, the chamber, besides paying a fine, had to let off firecrackers by way of apology.

Vagrant ways. A resolution passed in Liling county prohibited the drumming of New Year greetings or the chanting of praises to the local deities or the singing of lotus rhymes.[13] Various other counties have passed resolutions prohibiting this; in other places, these practices have disappeared of themselves, and no one engages in them anymore. The "beggar-bullies" or "vagabonds," who used to be extremely evil, now have no alternative but to submit to the peasant associations. In Shaoshan, Xiangtan county, the vagabonds used to make the temple of the Rain God their regular haunt and could not be persuaded by anyone, but since the rise of the associations they have all stolen away. The peasant association in Huti township in the same county caught three such vagabonds and made them carry clay for the brick kilns. Resolutions have been passed prohibiting the wasteful customs associated with New Year calls and gifts.

Besides these, a great many other minor prohibitions have been introduced in various places, such as the Liling prohibitions on incense-burning processions to propitiate the god of pestilence, on buying preserves and fruit for ritual presents, on burning ritual paper garments during the Festival of the Dead, and on pasting up good-luck posters at the New Year. At Gushui in Xiangxian county, there is even a prohibition on smoking water pipes. In the Second District, letting off firecrackers and ceremonial guns is forbidden, with a fine of 1.20 yuan for the former and 2.40 yuan for the latter. Religious rites for the dead are prohibited in the Seventh and Twentieth Districts. In the Eighteenth District, it is forbidden to make funeral gifts of money. Things like these, which defy enumeration, may be generally called "peasant bans and prohibitions." They are of great significance in two

[13]*lotus rhymes:* popular Buddhist chants

respects. First, they represent a revolt against bad customs, such as gaming, gambling, and opium smoking. These customs arose out of the rotten political environment of the landlord class and are swept away once its authority is overthrown. Second, the prohibitions are a form of self-defense against exploitation by city merchants; such are the prohibitions on feasts and on buying preserves and fruit for ritual presents. Because manufactured goods are extremely dear and agricultural products are extremely cheap, the peasants are very ruthlessly exploited by the merchants, and they must therefore engage in passive resistance. The reason for all this is that the unscrupulous merchants exploited them; it is not a matter of their rejecting manufactured goods in order to uphold the Doctrine of Oriental Culture.[14] The peasants' economic protection of themselves necessitates that the peasants organize consumers' cooperatives for collective sale and production. Furthermore, it is also necessary for the government to provide help to the peasant associations in establishing credit cooperatives. If these things were done, the peasants would naturally find it unnecessary to ban the outflow of grain as a method of keeping down the price; nor would they have to prohibit the inflow of manufactured goods as the sole method of economic self-defense.

10. Eliminating Banditry

In my opinion, no ruler in any dynasty from Yao, Shun, Yu, and Tang [in ancient times] down to the Qing emperors and the presidents of the Republic has ever shown as much prowess in eliminating banditry as have the peasant associations today. Wherever the peasant associations are powerful, there is not even the shadow of a bandit. It is truly amazing! In many places there are no longer even those pilferers who stole vegetables at night. Though there are still pilferers in some places, in the counties I visited, even including those that were formerly bandit-ridden, there was no trace of bandits. The reasons are: First, the members of the peasant associations are spread out everywhere over the hills and dales, spear or cudgel in hand, ready to go into action in their hundreds, so that the bandits have nowhere to hide. Second, since the peasants have prohibited the outflow of rice, the price of rice is extremely modest. It was six yuan a

---

[14]*Doctrine of Oriental Culture:* a belief that foreign (that is, Western) things are inferior to Chinese or Oriental things. Mao is making fun of the doctrine and claiming that peasants reject modern things not because of the doctrine but because merchants are overcharging for them.

picul[15] of rice last spring but only two yuan last winter. The poor peasants can buy more grain with less money. And the problem of food has become less serious than in the past for the people. Third, members of the secret societies have all joined the peasant associations, in which they can openly play the hero and vent their grievances, so that there is no further need for the secret "mountain," "lodge," "shrine," and "river" forms of organization. In killing the pigs and sheep of the local tyrants and evil gentry and imposing heavy levies and fines, they have adequate outlets for their feelings against those who oppressed them. Fourth, the armies are recruiting large numbers of soldiers and many of the "unruly" have joined up. Thus the evil of banditry has been eliminated with the rise of the peasant movement. On this point, even the well-to-do approve of the peasant associations. Their comment is:

> The peasant associations? Well, to be fair, there is also something to be said for them.

In prohibiting gaming, gambling, and opium smoking, and in eliminating banditry, the peasant associations have won general approval.

11. Abolishing Exorbitant Levies

As the whole country has not yet been unified and the authority of the imperialists and the warlords has not been overthrown, there is as yet no way of removing the heavy burden of government taxes and levies on the peasants or, more explicitly, of removing the burden of expenditure for the revolutionary army. However, the exorbitant levies imposed on the peasants when the local bullies and bad gentry dominated rural administration, for example, the surcharge on each *mu* of land, have been abolished or at least reduced with the rise of the peasant movement and the downfall of the local bullies and bad gentry. This too should be counted among the achievements of the peasant associations.

12. The Movement for Education

In China education has always been the exclusive preserve of the landlords, and the peasants have had no access to it. But the landlords' culture is completely created by the peasants, for its sole source

---

[15]*picul:* about 133 pounds

is the peasants' sweat and blood that they plundered. In China, more than 90 percent of the citizens have had no access to culture, and of these the overwhelming majority are peasants. The moment the power of the exploiting class was overthrown in the rural areas, the peasants' movement for education began. See how the peasants who hitherto detested the schools are today zealously setting up evening classes! They always disliked the "foreign-style school." When I was going to school and saw that the peasants were against the "foreign-style school," I, too, used to identify myself with the general run of "foreign-style students and teachers" and stand up for it, feeling always that the peasants were "stupid and detestable people." Only in the 14th year of the Republic [1925], when I lived in the country-side for half a year, did I realize that I had been wrong and the peasants' reasoning was extremely correct. The texts used in the rural primary schools were entirely about urban things and unsuited to rural needs. Besides, the attitude of the primary school teachers toward the peasants was very bad and, far from being helpful to the peasants, they came to be disliked by the peasants. Hence the peasants preferred the old-style schools (the so-called "Chinese classes") to the modern schools and the old-style teachers to the ones in the primary schools. Now the peasants are enthusiastically establishing evening classes, which they call "peasant schools." Some have already been opened, others are being organized, and on the average there is one school for every township peasant association. The peasants are very enthusiastic about setting up these evening schools and regard them, and only them, as truly their own. The sources of funds for the evening schools come from the local "public revenue from superstition," from ancestral temple funds, and from other idle public funds or property. The county education boards wanted to use this money to establish national primary schools (that is, "foreign-style schools" not suited to the needs of the peasants), while the peasants wanted to set up peasant schools. Inevitably, there were clashes between the two sides, and the result was generally that both got some of the money, though there were places where the peasants got it all. The development of the peasant movement has naturally resulted in raising their cultural level. Before long, several schools will have sprung up in the villages throughout the province; this is quite different from the empty talk about "universal education," which the intelligentsia and the so-called "educationalists" have been bandying back and forth and which after all this time remains an empty phrase.

13. The Cooperative Movement

The peasants really need cooperatives, especially consumers', marketing, and credit cooperatives. When they buy goods, the merchants exploit them; when they sell their farm produce, the merchants cheat them; when they borrow money or rice, they are fleeced by the usurers; and they are eager to find a solution to these three problems. During the fighting in the Yangtse valley last winter, when trade routes were cut and the price of salt went up in Hunan, a great many peasants organized cooperatives for salt. When the landlords deliberately stopped lending, there were many attempts by the peasants to organize credit agencies because they needed to borrow money. A major problem is the absence of detailed, standard rules of organization. In all localities, many of these spontaneously organized peasant cooperatives fail to conform to cooperative principles; as a result, the comrades engaged in the peasant movement are always eagerly enquiring about "rules and regulations." Given proper guidance, the cooperative movement can spread everywhere along with the growth of the peasant associations. Because the term *hezuo* is not at all familiar to the peasants, [the idea] could also be rendered as *hehuopu.**

14. Building Roads and Embankments

This, too, is one of the achievements of the peasant associations. Before there were peasant associations the roads in the countryside were terrible. Because roads cannot be repaired without money, and the wealthy were unwilling to dip into their purses, the roads were left in bad shape. If there was any road work done at all, it was done as an act of charity; a little money was collected from families "wishing to gain merit in the next world," and a few narrow, skimpily paved roads were built. With the rise of the peasant associations, orders have been given specifying the required width—three, five, seven, or ten feet, according to the requirements of the different routes—and each landlord along a road has been ordered to build a section. Once the order is given, who dares to disobey? In a short time many good roads have appeared. This is no work of charity but the result of compulsion, and

---

Hezuo (cooperate, literally "work together"), and *hezuoshe* (cooperative) have been the standard Chinese terms since the 1920s. The alternative that Mao suggests, *hehuopu*, means literally joint goods shop." It is in fact this coinage which he used for "cooperative" in the title of this section of his report in the original version.

a little compulsion of this kind is not at all a bad thing. The same is true of the embankments. The ruthless landlords were always out to take what they could from the tenant-peasants and would never spend even a few copper cash on embankment repairs; they would leave them to dry up and the tenant-peasants to starve, caring about nothing but the rent. Now that there are peasant associations, they can be bluntly ordered to repair the embankments. When a landlord refuses, the association will tell him very affably:

"Very well! If you won't do the repairs, you will contribute grain, a *dou*[16] for each workday." As this is a bad bargain for the landlord, he hastens to do the repairs. Consequently many defective embankments have been turned into good ones.

The fourteen deeds enumerated above have all been accomplished by the peasants under the command of the peasant associations; would the reader please consider and say whether any of them is bad? Only the local bullies and bad gentry, I think, will call them bad. Curiously enough, it is reported from Nanchang that Chiang Kaishek, Zhang Jingjiang, and other such gentlemen do not altogether approve of the activities of the Hunan peasants. This opinion is shared by Liu Yuezhi and other right-wing leaders in Hunan, all of whom say, "They have simply gone Red." But where would the national revolution be without this bit of Red? To talk about arousing the masses of the people day in and day out and then to be scared to death when the masses do rise—what difference is there between this and Lord She's love of dragons?*

---

[16]dou: thirty pounds

*The reference is to an anecdote in the *Xin xu* (New Prefaces) of Liu Xiang (76–5 B.C.E.), a descendant of Liu Bang. Lord She professed such a love of dragons that he decorated his whole palace with drawings and carvings of them. Pleased by this report, a real dragon paid him a visit and frightened Lord She out of his wits.

2

On New Democracy

January 15, 1940

In this essay, Mao begins his effort to organize and categorize the experience of China over the past generation by casting it in a broader narrative of imperialist aggression and "feudal" collaboration with invaders going back a century to the Opium War of 1839–42. "On New Democracy" seeks nothing less than to create a meaningful "history" out of "experience," to tell the national story of the new China for a broader public in 1940. The telling, naturally, puts the CCP at the center of the story as the savior that can bring the nation back together and lead it to a prosperous and just future. Mao convinced both inner-party audiences of cadres and the broader public of students, urban professionals, independent political activists, and enough of China's rural population to gain popular support for the CCP in the 1940s as it fought against the GMD (Nationalist party) for the right to rule China.

The essay was first published in January 1940 in a small party journal in Yan'an. Through the Rectification Movement of 1942–44, Mao's ideas became the dominant philosophy of the CCP, and "On New Democracy" contains the core of those ideas. The main audience for Mao's "story" were educated Chinese, who would be needed to staff the new state run by the CCP. This massive expansion of personnel required a coherent set of goals. Just as Mao had tried to claim the mantle of Sun Yat-sen's 1911 Revolution in his 1927 "Report on the Peasant Movement in Hunan," (see Document 1), so did he seek to capture the spirit of the most important intellectual event of his generation—the May Fourth Movement or New Culture Movement—in this essay. Indeed, his description of the events around the famous 1919 student demonstration and associated cultural trends is so detailed and coherent that Western scholars long accepted it as completely accurate.

"Lun xin minzhuzhuyi," in *Mao Zedong ji*, ed. Takeuchi Minoru (Tokyo: Hokobosha, 1970), 7:143–202, which is taken from the 1944 Chinese edition of Mao's *Selected Works*. This translation by Stuart Schram in Stuart R. Schram and Nancy J. Hodes, eds., *Mao's Road to Power: Revolutionary Writings, 1912–1949: Vol. VII, 1940–1941* (Armonk, N.Y.: M. E. Sharpe, forthcoming). I have edited the text very lightly to minimize technical terms. I have not included Schram's extensive notations of textual variations between this original and the post-1949 editions of Mao's *Selected Works*.

The essay is organized as a lecture, explaining what New Democracy is and why it is the right path for China in a series of questions and answers. Throughout the essay, Mao assumes that democracy means a government that reflects the interests of China's ordinary people; he does not mean elected representative government in the American sense. Along the way, Mao gives a detailed review of modern Chinese history and its stages of development. The turning point from "old democratic revolution" to "new democratic revolution" was the May Fourth Movement. In Mao's version of the story, the shift makes sense in Marxist terms—from bourgeois leadership of the old revolution, which achieved national independence in 1911, to proletarian (that is, CCP) leadership of the new revolution, which was to achieve socialism in the future. He adapts Sun Yat-sen's "Three People's Principles" (nationalism, democracy, and people's livelihood) by adding the "Three Great Policies"— alliance with Soviet Russia, cooperation with the CCP, and attention to the needs of workers and peasants. Mao appeals to his non-CCP audience by putting the party's "maximum program" (communism) into the more distant future and guaranteeing a long period of the "minimum program" (a democratic alliance of all nationalist parties).

Two of Mao's most famous lines appear in this essay. The first, "The history of modern China is a history of imperialist aggression," summarizes the broad appeal of Mao's leadership across China. He successfully casts himself and the CCP as nationalists and anti-imperialists. In this the CCP was true to its word, ejecting foreigners and nationalizing foreign industries after 1949. Mao's second famous quotation reflects his militant nature: "There is no construction without destruction." Yet this comment is made in reference not to social or political order but to culture and ideology (it would later be used to justify social violence in the Cultural Revolution). Also, it is important to note that these words evoke the ancient Chinese philosophy of yin and yang (flowing and damming; motion and rest), indicating the fit between the Marxist philosophy of dialectics and the traditional Chinese worldview. Mao's story gave the Chinese a way of understanding their place in the world and told them what to do about it.

I. WHITHER CHINA?

A lively atmosphere has prevailed throughout the country ever since the War of Resistance[1] began. There is a general feeling that a way out of the impasse has been found, and people no longer knit their brows in despair. Of late, however, the dust and din of compromise and anti-Communism have once again filled the air, and once again the people are thrown into bewilderment. Most susceptible, and the first to be affected, are the intellectuals and the young students. So the questions "What is to be done?" and "Where is China headed?" have once again arisen. On the occasion of the publication of *Chinese Culture*,[2] it may therefore be profitable to clarify the political and cultural trends in China. I am a layman in matters of culture; I would like to study them, but have only just begun to do so. Fortunately, there are many comrades in Yan'an who have written at length in this field, so that my rough and ready words may serve the same purpose as the beating of the gongs before a theatrical performance. Our observations may contain a grain of truth for the nation's advanced cultural workers and may serve as a modest spur to induce them to come forward with valuable contributions of their own. We hope that they will join in the discussion to reach correct conclusions which will meet our national needs. To "seek truth from facts"[3] is the scientific approach, and presumptuously to claim infallibility and lecture people will *assuredly* never settle anything. The disaster that has befallen our nation is extremely grave, and only a scientific approach and a spirit of responsibility can lead it on to the road of liberation. There is but one truth, and the question of whether or not one has it depends not on subjective boasting but on objective practice. The only yardstick of truth is the revolutionary practice of millions of people. This, I think, can be regarded as the attitude of *Chinese Culture*.

---

[1] *War of Resistance:* China's war against Japan, which began with Japan's invasion in July 1937 and ended with the end of World War II in August 1945

[2] Chinese Culture: the journal in Yan'an in which Mao first published this essay

[3] *seek truth from facts:* one of the most famous phrases in Chinese communism. This credo of one school of Neo-Confucianism was carved over the door of Mao's high school. The phrase is especially famous for its use in the early post-Mao period of the late 1970s to reverse the excesses of the Cultural Revolution.

II. WE WANT TO BUILD A NEW CHINA

For many years we Communists have struggled not only for a political and economic revolution, but for a cultural revolution as well. The goal of all these revolutions is to build a new society and a new state for the Chinese nation. That new society and new state will have not only a new politics and a new economy but a new culture. In other words, not only do we want to change a China that is politically oppressed and economically exploited into a China that is politically free and economically prosperous, we also want to change the China which is being kept ignorant and backward under the sway of the old culture into an enlightened and progressive China under the sway of a new culture. In short, we want to build a new China. Our aim in the cultural sphere is to build a new Chinese national culture.

III. CHINA'S HISTORICAL CHARACTERISTICS

We want to build a new national culture, but what kind of culture should it be?

Any given culture (as an ideological form) is a reflection of the politics and economics of a given society, and the former in turn has a tremendous influence upon the latter; politics is the concentrated expression of economics. This is our fundamental view of the relation of culture to politics and economics and of the relation of politics to economics. It follows that the form of culture is first determined by the political and economic form, and only then does it influence the given political and economic form. Marx says, "It is not the consciousness of men that determines their being, but, on the contrary, their social being that determines their consciousness."[4] He also says, "The philosophers have only interpreted the world in various ways; the point, however, is to change it."[5] For the first time in human history, these scientific formulations correctly solved the problem of the relationship between consciousness and existence, and they are the basic points of departure underlying the dynamic revolutionary theory of knowledge as the reflection of reality which was later elaborated so

[4]Karl Marx, "Preface to *A Contribution to the Critique of Political Economy*," in *Selected Works of Marx and Engels*, English ed. (Moscow: Foreign Languages Publishing House, 1958), 1:363.
[5]Karl Marx, "Theses on Feuerbach," in *Selected Works of Marx and Engels*, 2:405.

profoundly by Lenin. These basic points of departure must be kept in mind in our discussion of China's cultural problems.

Thus it is quite clear that the old national culture we want to eliminate is inseparable from the old national politics and economics, while the new national culture which we want to build up is inseparable from the new national politics and economics. The old politics and economics of the Chinese nation form the basis of its old culture, just as its new politics and economics will form the basis of its new culture.

What are China's old politics and economics? And what is her old culture?

From the Zhou and Qin dynasties[6] onward, Chinese society was feudal, as were its politics and its economy. And the culture, reflecting the politics and economy, was feudal culture.

Since the invasion of foreign capitalism and the gradual growth of capitalist elements in Chinese society, that is, during the hundred years from the Opium War to the Sino-Japanese War, the country has changed by degrees into a colonial, semi-colonial, and semi-feudal society. China today is colonial in the enemy-occupied areas and basically semi-colonial in the non-occupied areas, and it is predominantly feudal in both. Such, then, is the character of present-day Chinese society and the state of affairs in our country. The politics and the economy of this society are predominantly colonial, semi-colonial, and semi-feudal, and the culture, reflecting the politics and economy, is also colonial, semi-colonial, and semi-feudal.

It is precisely against these predominant political, economic, and cultural forms that our revolution is directed. What we want to get rid of is the old colonial, semi-colonial, and semi-feudal politics and economy and the old culture. And what we want to build up is their direct opposite, i.e., the new politics, the new economy, and the new culture of the Chinese nation.

What, then, are the new politics and the new economy of the Chinese nation, and what is its new culture?

In the course of its history the Chinese revolution must go through two stages, first, the democratic revolution, and second, the socialist revolution, and by their very nature they are two different revolutionary processes. But what I call democracy no longer belongs to the old category, it is not the old democracy; it belongs to the new category, it is New Democracy.

---

[6]The Zhou and Qin dynasties ruled from the eleventh to the third centuries B.C.E., setting the model for later Chinese empires.

It can thus be affirmed that China's new politics are the politics of New Democracy, that China's new economy is the economy of New Democracy, and that China's new culture is the culture of New Democracy.

Such are the historical characteristics of the Chinese revolution at the present time. Any political party, group, or person taking part in the Chinese revolution that fails to understand this will not be able to direct the revolution and lead it to victory, but will be cast aside by the people and left to grieve out in the cold.

IV. THE CHINESE REVOLUTION IS PART OF THE WORLD REVOLUTION

The historical characteristic of the Chinese revolution lies in its division into the two stages, democracy and socialism, the first being no longer democracy in general, but democracy of the Chinese type, a new and special type, namely, New Democracy. How, then, has this historical characteristic come into being? Has it been in existence for the past hundred years, or is it of recent origin?

A brief study of the historical development of China and of the world shows that this characteristic did not emerge immediately after the Opium War, but took shape later, after the First Imperialist World War and the October Revolution[7] in Russia. Let us now examine the process of its formation.

Clearly, it follows from the colonial, semi-colonial, and semi-feudal character of present-day Chinese society that the Chinese revolution must be divided into two stages. The first step is to change the colonial, semi-colonial, and semi-feudal form of society into an independent, democratic society. The second is to carry the revolution forward and build a socialist society. At present the Chinese revolution is taking the first step.

The first step began with the Opium War in 1840, that is to say, when China's feudal society started changing into a semi-colonial and semi-feudal one. Then came the Movement of the Taiping Heavenly Kingdom, the Coup of 1898,* the Sino-French War, the Sino-Japanese

---

[7]The October Revolution in 1917 brought Lenin and the Bolsheviks to power in Russia and led to the formation of the USSR.

*Here, as in his speech of May 4, 1939, Mao refers to the events of 1898 as a "coup." Obviously he was thinking of the suppression of the Reform Movement, rather than of that movement itself. In the *Selected Works* version, this event has been moved to its

War, the Revolution of 1911, the May Fourth Movement, the May Thirtieth Movement, the Northern Expedition, the Agrarian Revolution, the December Ninth Movement, and the present War of Resistance against Japan.[8] Together these have taken up a whole century and in a sense they represent that first step, being struggles waged by the Chinese people, on different occasions and in varying degrees, against imperialism and the feudal forces in order to build up an independent, democratic society and complete the first revolution. The Revolution of 1911 was in a fuller sense the beginning of that revolution. In its social character, this revolution is a bourgeois-democratic and not a proletarian-socialist revolution. It is still unfinished and still demands great efforts, because to this day its enemies are still very strong. When Mr. Sun Yat-sen said, "The revolution is not yet completed, all my comrades must struggle on," he was referring to the bourgeois-democratic revolution.

A change, however, occurred in China's bourgeois-democratic revolution after the outbreak of the First Imperialist World War in 1914 and the founding of a socialist state on one-sixth of the globe as a result of the Russian October Revolution of 1917.

Before these events, the Chinese bourgeois-democratic revolution came within the old category of the bourgeois-democratic world revolution, of which it was a part.

Since these events, the Chinese bourgeois-democratic revolution has changed, it has come within the new category of bourgeois-democratic revolutions and, as far as the alignment of revolutionary forces is concerned, forms part of the proletarian-socialist world revolution.

Why? Because the First Imperialist World War and the first victorious socialist revolution, the October Revolution, have changed the whole course of world history and ushered in a new era.

In an era in which the world capitalist front has collapsed in one corner of the globe (a corner which occupies one-sixth of the world's surface), and has fully revealed its decadence everywhere else, in an era in which the remaining capitalist portions cannot survive without relying more than ever on the colonies and semi-colonies, in an era in which a socialist state has been established and has proclaimed its readiness to fight in support of the liberation movement of all colonies and semi-colonies, and in which the proletariat of the capitalist coun-

---

proper chronological place, after the Sino-French and Sino-Japanese wars, but it remains a coup.

[8]This list of events, from the Opium War to the War of Resistance, reflects the major events in Chinese foreign relations from 1839 to 1940.

tries is steadily freeing itself from the social-imperialist influence of the social-democratic parties and has proclaimed its support for the liberation movement in the colonies and semi-colonies—in such an era, a revolution in any colony or semi-colony that is directed against imperialism, that is to say, against the international bourgeoisie and international capitalism, no longer comes within the old category of the bourgeois-democratic world revolution, but within the new category. It is no longer part of the old bourgeois and capitalist world revolution, but is part of the new world revolution, the proletarian-socialist world revolution. Such revolutionary colonies and semi-colonies can no longer be regarded as allies of the counter-revolutionary front of world capitalism; they have become allies of the revolutionary front of world socialism.

Although during its first stage or step, such a revolution in a colonial and semi-colonial country is still fundamentally bourgeois-democratic in its social character, and although its objective demand is still basically to clear the path for the development of capitalism, it is no longer a revolution of the old type, led entirely by the bourgeoisie, with the aim of establishing a capitalist society and a state under bourgeois dictatorship. It is rather a revolution of the new type, in which the proletariat participates in or exercises the leadership, and having as its aim, in the first stage, the establishment of a new-democratic society and a state under the joint dictatorship of all the revolutionary classes. In the course of its progress, there may be a number of further sub-stages, because of changes on the enemy's side and within the ranks of our allies, but the fundamental character of the revolution remains unchanged and it will remain the same until the time of the socialist revolution.

Such a revolution attacks imperialism at its very roots, and is therefore not acceptable to imperialism, which on the contrary opposes it. But it is acceptable to socialism, and is supported by the land of socialism[9] and by the international socialist proletariat.

Therefore, such a revolution cannot but become part of the proletarian-socialist world revolution.

The correct thesis that "the Chinese revolution is part of the world revolution" was put forward as early as 1924–27 during the period of China's Great Revolution.[10] It was put forward by the Chinese Communists and endorsed by all those taking part in the anti-imperialist

[9]*land of socialism:* the Soviet Union
[10]*China's Great Revolution:* another name for the Northern Expedition of 1926–27, which brought GMD rule to central China and the establishment of the capital of the Republic of China in Nanjing

and anti-feudal struggle of the time. At that time, however, the significance of this thesis was not fully expounded, and consequently it was only vaguely understood. I remember that during his eastern expedition against Cheng Jiongming in 1925, Mr. Chiang Kai-shek made a speech on reaching Chao Shan,* in which he also said, "China's revolution is part of the world revolution."

The "world revolution" no longer refers to the old world revolution, for the old bourgeois world revolution has long been a thing of the past; it refers to the new world revolution, the socialist world revolution. Similarly, to form "part of" means to form part not of the old bourgeois revolution, but of the new socialist revolution. This is a tremendous change unparalleled in the history of China and of the world.

The Chinese Communists put forward this correct thesis on the basis of Stalin's theory.

As early as 1918, in an article commemorating the first anniversary of the October Revolution, Stalin wrote:

> The great world-wide significance of the October Revolution chiefly consists of the following three points. First, it has widened the scope of the national question and converted it from the particular question of national oppression to the general question of emancipating the oppressed peoples, colonies, and semi-colonies from imperialism. Second, it has opened up wide possibilities for their emancipation and the right paths towards it, has thereby greatly facilitated the cause of the emancipation of the oppressed peoples of the West and the East, and has drawn them into the common path of the victorious struggle against imperialism. Third, it has thereby erected a bridge between the socialist West and the enslaved East, thus creating a new front of revolutions against world imperialism, extending from the proletarians of the West, through the Russian Revolution, to the oppressed peoples of the East. (Zhang Zhongshi, *Stalin on the National Question*, Liberation Press, Yan'an, p. 130.)[11]

Since writing this article, Stalin has again and again expounded the theory that revolutions in the colonies and semi-colonies have broken away from the old category and become part of the proletarian-socialist revolution. The clearest and most precise explanation is given in an article published on June 30, 1925, in which Stalin carried on a

---

*The reference is to Chaoxian and Shantou, two small cities in the northeastern corner of Guangdong, or to the short railroad connecting them.

[11]Translated in J. V. Stalin, "The October Revolution and the National Question," in *Works*, English ed. (Moscow: Foreign Languages Publishing House, 1953), 4:169–70.

controversy with the Yugoslav nationalists of the time. Entitled "The National Question Once Again," it is included in a book translated by Chang Zhongshi and published under the title *Stalin on the National Question.* It contains the following passage:

> Comrade Semich refers to a passage in Stalin's book *Marxism and the National Question,* written at the end of 1912. There it says that "the national struggle is a struggle of the bourgeois classes among themselves." Evidently, by this Semich is trying to suggest that his formula defining the social significance of the national movement under the present historical conditions is correct. But Stalin's pamphlet was written before the imperialist war, when the national question was not yet regarded by Marxists as a question of world significance, when the Marxists' fundamental demand for the right to self-determination was regarded not as part of the proletarian socialist revolution, but as part of the bourgeois-democratic revolution. It would be ridiculous not to see that since then the international situation has radically changed, that the war in Europe, on the one hand, and the October Revolution in Russia, on the other, transformed the national question from a part of the bourgeois-democratic revolution into a part of the proletarian-socialist revolution. As far back as October 1916, in his article, "The Discussion on Self-Determination Summed Up," Lenin said that the main point of the national question, the right to self-determination, had ceased to be a part of the general democratic movement, that it had already become a component part of the general proletarian, socialist revolution. I do not even mention many other profound works on the national question by Lenin and by other representatives of Russian communism. What significance can Semich's reference to the passage in Stalin's book, written in the period of the bourgeois-democratic revolution in Russia, have at the present time, when as a consequence of the new historical situation, we have entered a new epoch, the present epoch of world proletarian revolution? It can only signify that Comrade Semich completely quotes outside of space and time, without reference to the living historical situation, and thereby violates the elementary requirements of dialectics, and ignores the saying that what is right for one historical situation may prove to be wrong in another historical situation.[12]

From this it can be seen that there are two kinds of world revolutions. The first belongs to the bourgeois or capitalist category. The era of this kind of world revolution is long past; it came to an end as far

[12]Translated in J. V. Stalin, "The National Question Once Again," in *Works,* English ed. (Moscow: Foreign Languages Publishing House, 1954), 7:225–27.

back as 1914, when the First Imperialist World War broke out, and above all in 1917, when the Russian October Revolution took place. The second kind, namely the proletarian-socialist world revolution, thereupon began. This type of revolution has the proletariat of the capitalist countries as its main force, and the oppressed peoples of the colonies and semi-colonies as its allies. No matter what classes, parties, or individuals in an oppressed nation join the revolution, and no matter whether they are conscious of this point or understand it subjectively, so long as they oppose imperialism, their revolution becomes part of the proletarian-socialist world revolution and they become its allies.

Today, the Chinese revolution has taken on still greater significance. This is a time when the economic and political crises of capitalism are dragging the world more and more deeply into the Second Imperialist War,[13] when the Soviet Union has reached the period of transition from socialism to communism and is capable of leading and helping the proletariat, oppressed nations, and all revolutionary people of the whole world in their fight against imperialist war and capitalist reaction, when the proletariat of the capitalist countries is preparing to overthrow capitalism and establish socialism, and when the proletariat, the peasantry, the intellectuals, and the petty bourgeoisie in China have become a mighty independent political force under the leadership of the Chinese Communist Party. Situated as we are in this day and age, should we not make the appraisal that the Chinese revolution has taken on still greater world significance? I think we should. The Chinese revolution is a great part of the world revolution.

Although the Chinese revolution in this first stage (with its many sub-stages) is a new type of bourgeois-democratic revolution and is not yet itself the newest type of proletarian-socialist revolution in its social character, it has long become a part of the proletarian-socialist world revolution and now even a very important part and a great ally of this world revolution. The first step or stage in our revolution is definitely not, and cannot be, the establishment of a capitalist society under the dictatorship of the Chinese bourgeoisie, but will result in the establishment of a new-democratic society under the joint dictatorship of all the revolutionary classes of China. The revolution will then be carried forward to the second stage, in which a socialist society will be established in China.

[13] *Second Imperialist War:* World War II

This is the fundamental characteristic of the Chinese revolution of today, of the new revolutionary process of the past twenty years (counting from the May Fourth Movement), and its concrete living essence.

V. THE POLITICS OF NEW DEMOCRACY

The new historical characteristic of the Chinese revolution is its division into two stages, the first being the new-democratic revolution. How does this manifest itself concretely in internal political and economic relations? Let us consider the question.

Before the May Fourth Movement of 1919 (which occurred after the First Great Imperialist War of 1914 and the Russian October Revolution of 1917), the petty bourgeoisie and the bourgeoisie (through their intellectuals) were the political leaders of the bourgeois-democratic revolution. The Chinese proletariat had not yet appeared on the political scene as an awakened and independent class force, but participated in the revolution only as a follower of the petty bourgeoisie and the bourgeoisie. Such was the case with the proletariat at the time of the Revolution of 1911.

After the May Fourth Movement, the chief political leader of China's bourgeois-democratic revolution was no longer the single class of the bourgeoisie, and the proletariat also participated in the political leadership. The Chinese proletariat rapidly became an awakened and independent political force as a result of its maturing and of the influence of the Russian Revolution. It was the Chinese Communist Party that put forward the slogan "Down with imperialism" and the thoroughgoing program for the whole of the bourgeois-democratic revolution, and it was the Chinese Communist Party alone that carried out the Agrarian Revolution.

Because the Chinese bourgeoisie is the bourgeoisie of a colonial and semi-colonial country, and because it is oppressed by imperialism, it retains at certain periods and to a certain degree—even in the era of imperialism—a certain revolutionary nature which leads it to fight against foreign imperialism and the domestic governments of bureaucrats and warlords (instances of opposition to the latter can be found in the periods of the Revolution of 1911 and the Northern Expedition, that is to say at periods when the bourgeoisie itself did not exercise power). It can ally itself with the proletariat and the petty bourgeoisie against such enemies as it is ready to oppose. In this respect the Chinese bourgeoisie differs from the bourgeoisie of the old Russian empire. Since the old Russian empire was itself already a military-feudal imperialism which carried on aggression against other countries, the Russian bourgeoisie was entirely lacking in revolutionary quality. There, the task of the proletariat was to oppose the bourgeoisie, not to unite with it. But because China is a colonial and semi-colonial country which is a victim of aggression, the Chinese bourgeoisie has a revolutionary quality at certain periods and to a certain degree. Here, the task of the proletariat is not to neglect this revolutionary quality of the bourgeoisie, or the possibility of establishing a united front with it against imperialism and the bureaucrat and warlord governments.

At the same time, however, precisely because the Chinese bourgeoisie is the bourgeoisie of a colonial and semi-colonial country, it is extremely flabby economically and politically, and it also has another quality, namely a proneness to compromise with the enemies of the revolution. The Chinese bourgeoisie, and especially the big bourgeoisie, even when it takes part in the revolution, is unwilling to break with imperialism completely, and is, moreover, closely associated with exploitation through the land in the rural areas. Thus it is neither willing nor able to overthrow imperialism thoroughly, still less to overthrow the feudal forces thoroughly. So neither of the two basic problems or tasks of China's bourgeois-democratic revolution can be solved or accomplished by the bourgeoisie. During the long period between 1927 and 1936, it nestled in the arms of the imperialists, formed an alliance with the feudal forces, betrayed its own revolutionary programs, and opposed the revolutionary people of the time.[14] During the War of Resistance, the section of the big bourgeoisie represented by Wang Jingwei[15] has once again capitulated to the enemy. This constitutes a fresh betrayal on the part of the big bourgeoisie. This is also a point with respect to which the bourgeoisie in China differs from the earlier bourgeoisie of the advanced countries in Europe and America, especially France. When the European and American countries were still in their revolutionary era, the bourgeoisie of those countries, and especially of France, was comparatively thorough in

[14]Mao is describing and criticizing the actions of the Nationalist government under Chiang Kai-shek.
[15]*Wang Jingwei:* a leader of the Guomindang who was sidelined by Chiang Kai-shek and then became head of the collaborationist regime in Nanjing during the Sino-Japanese War

carrying out the revolution. In China, the bourgeoisie does not possess even this degree of thoroughness.

On the one hand revolutionary nature, and on the other hand proneness to conciliation—such is the dual character of the Chinese bourgeoisie, which faces both ways. Even the bourgeoisie in European and American history shared this dual character. When confronted by a formidable enemy, they united with the workers and peasants against him, but when the workers and peasants awakened, they turned round to unite with the enemy against the workers and peasants. This is a general rule applicable to the bourgeoisie everywhere in the world, but the trait is more pronounced in the Chinese bourgeoisie.

In China, the situation is extremely clear. Whoever can lead the people in overthrowing imperialism and the feudal forces will be able to win the people's confidence, for the mortal enemies of the people are imperialism and the feudal forces, especially imperialism. Today, whoever can lead the people in driving out Japanese imperialism and introducing democratic government will be the savior of the people. If the Chinese bourgeoisie can fulfill this responsibility, no one will be able to refuse his admiration; but if it cannot do so, the responsibility will inevitably fall upon the shoulders of the proletariat.

Therefore, the proletariat, the peasantry, the intellectuals, and the other sections of the petty bourgeoisie undoubtedly constitute the basic forces determining China's fate. These classes, some already awakened and others in the process of awakening, will necessarily become the basic components of the state and governmental structure in the democratic republic of China. The Chinese democratic republic which we now desire to establish can only be a democratic republic under the joint dictatorship of all the anti-imperialist and anti-feudal people. That is to say that it will be a new-democratic republic, a republic of the genuinely revolutionary new Three People's Principles with their Three Great Policies.

This new-democratic republic will be different from the old European-American form of capitalist republic under bourgeois dictatorship, which is the old democratic form and already out-of-date. On the other hand, it will also be different from the socialist republic of the newest Soviet type under the dictatorship of the proletariat which is already flourishing in the USSR, and which, moreover, will be established in all the capitalist countries and will undoubtedly become the dominant form of state and governmental structure in all the advanced countries. For a certain historical period, however, this form is not suitable for the colonial and semi-colonial countries. During this period, therefore, a third form of state must be adopted in all colonial and semi-colonial countries, namely, the new-democratic republic. This is a form suited to a certain historical period, and is therefore a transitional form; nevertheless, it is a form which is necessary and cannot be dispensed with.

Thus the multifarious types of state system in the world, classified according to their social character, can be reduced to three basic kinds: (1) republics under bourgeois dictatorship; (2) republics under the dictatorship of the proletariat; and (3) republics under the joint dictatorship of several revolutionary classes.

The first kind comprises the old democratic states. Today, after outbreak of the second imperialist war, there is already not the slightest trace of democracy in any of the capitalist countries. They have all been transformed, or are about to be transformed, into bloody military dictatorships of the bourgeoisie. Certain countries under the joint dictatorship of the landlords and the bourgeoisie can be grouped with this kind.

Apart from the Soviet Union, the second kind is ripening in capitalist countries, and in the future, it will be the dominant form throughout the world for a certain period.

The third kind is the transitional form of state in the revolutionary colonies and semi-colonies. To be sure, the various colonies and semi-colonies will necessarily have different characteristics, but these are only minor differences within the general framework of uniformity. So long as they are revolutionary colonial or semi-colonial countries, their state and governmental structure will of necessity be basically the same, namely, a new-democratic state under the joint dictatorship of several anti-imperialist classes. In China today, the new-democratic state takes the form of the Anti-Japanese United Front. It is anti-Japanese and anti-imperialist; it is also a united front, an alliance of several revolutionary classes. But unfortunately, despite the fact that the War of Resistance has been going on for so long, the work of democratizing the state has hardly started, and the Japanese imperialists have exploited this fundamental weakness to stride into our country. If nothing is done about it, our national future will be gravely imperiled. We hope that the Movement for Constitutional Government that has just started will prevent this danger.

The question under discussion here is that of the "state system." After several decades of wrangling since the last years of the Qing dynasty, it has still not been cleared up. Actually it is simply a question of the status of the various social classes within the state. The bourgeoisie, as a rule, conceals the problem of class status and carries out its one-class dictatorship under the "national" label. Such concealment is of no advantage to the revolutionary people and the matter should be clearly explained to them. The term "national" can be used, but the people of the nation do not include counterrevolutionaries and Chinese traitors, and are comprised of all revolutionary people. The kind of state we need today is a dictatorship of all the revolutionary classes over the counterrevolutionaries and Chinese traitors.

The so-called democratic system in modern states is usually monopolized by the bourgeoisie and has become simply an instrument for oppressing the common people. On the other hand, the Guomindang's Principle of Democracy means a democratic system shared by all the common people and not privately owned by the few. Such was the solemn declaration made in the Manifesto of the First National Congress of the Guomindang, held in 1924. For sixteen years the Guomindang has violated this declaration and as a result it has created the present grave national crisis. This is a gross blunder, which we hope the Guomindang will correct in the cleansing flames of resistance to Japan.

As for the question of "political power," this is a matter of how political power is organized, the form in which one social class or another chooses to arrange its apparatus of political power to oppose its enemies and protect itself. There is no state which does not have an appropriate apparatus of political power to represent it. China may now adopt a system of congresses, from the national congress down to the provincial, county, district, and township congresses, with all levels electing their respective governmental bodies. But if there is to be proper representation for each revolutionary class according to its status in the state, a proper expression of the people's will, a proper direction for revolutionary struggles, and a proper manifestation of the spirit of New Democracy, then a system of really universal and equal suffrage, irrespective of sex, creed, property, or education, must be introduced. Such is the system of democratic centralism. Only a government based on democratic centralism can fully express the will of all the revolutionary people and fight the enemies of the revolution most effectively. There must be a spirit of refusal to be "privately owned by the few" in the government and the army; without a genuinely democratic system this cannot be attained and the system of government and the state system will be out of harmony.

The state system, a joint dictatorship of all the revolutionary classes, and the system of government, democratic centralism—these constitute the politics of New Democracy, the republic of New Democracy, the republic of the Anti-Japanese United Front, the republic of the new Three People's Principles with their Three Great Policies, the Republic of China in reality as well as in name. Today we have a Republic of China in name but not in reality, and our present task is to create the reality that will fit the name.

Such are the internal political relations which a revolutionary China, a China fighting Japanese aggression, should and must establish without fail; such is the orientation, the only correct orientation, for our present work of national reconstruction.

VI. THE ECONOMY OF NEW DEMOCRACY

If such a republic is to be established in China, it must be new-democratic not only in its politics but also in its economy.

The big banks and the big industrial and commercial enterprises will become state-owned.

> Enterprises such as banks, railways, and airlines, whether Chinese-owned or foreign-owned, which are either monopolistic in character or too big for private management, shall be operated and administered by the state, so that private capital cannot dominate the livelihood of the people: this is the main principle of the regulation of capital.

This is another solemn declaration in the Manifesto of the Guomindang's First National Congress, and it is the correct policy for the economic structure of the new-democratic republic. But at the same time the republic will neither confiscate capitalist private property in general nor forbid the development of such capitalist production as does not "dominate the livelihood of the people," for China's economy is still very backward.

The republic will take certain necessary steps to confiscate the land of the big landlords and distribute it to those peasants having little or no land, carry out Mr. Sun Yat-sen's slogan of "land to the tiller," abolish feudal relations in the rural areas, and turn the land over to the private ownership of the peasants without establishing a socialist agriculture. A rich peasant economy will be allowed in the rural areas. Such is the policy of "equalization of land ownership." "Land to the tiller" is the correct slogan for this policy.

China's economy must develop along the path of the "regulation of capital" and the "equalization of landownership," and must never be "privately owned by the few"; we must never permit the few capitalists and landlords to "dominate the livelihood of the people"; we must never establish a capitalist society of the European-American type or allow the old semi-feudal society to survive. Whoever dares to go counter to this line of advance will certainly not succeed but will run into a brick wall.

Such are the internal economic relations which a revolutionary China, a China fighting Japanese aggression, must and necessarily will establish.

Such is the economy of New Democracy.

And the politics of New Democracy are the concentrated expression of the economy of New Democracy.

VII. REFUTATION OF BOURGEOIS DICTATORSHIP

More than 90 percent of the people are in favor of a republic of this kind with its new-democratic politics and new-democratic economy; "without such a republic, nothing can be achieved, for it accords with the natural principles and people's sentiments, goes with the trend of the world, meets the demands of the people, and has been pursued resolutely by people of foresight" (Sun Yat-sen's words). There is no alternative road.

What about the road to a capitalist society under bourgeois dictatorship? To be sure, that was the old road taken by the European and American bourgeoisie, but whether one likes it or not, neither the international nor the domestic situation allows China to do the same.

Judging by the international situation, that road is blocked. In its fundamentals, the present international situation is one of a struggle between capitalism and socialism, in which capitalism is on the downgrade and socialism on the upgrade. In the first place international capitalism, or imperialism, will not permit it. Indeed the history of modern China is a history of imperialist aggression, of imperialist opposition to China's independence and to her development of capitalism. Earlier revolutions failed in China because imperialism strangled them, and innumerable revolutionary martyrs died, bitterly lamenting the non-fulfillment of their mission. Today a powerful Japanese imperialism is forcing its way into China and wants to reduce her to a colony; it is not China that is developing Chinese capitalism but Japan that is developing Japanese capitalism in our country; and it is not the Chinese bourgeoisie but the Japanese bourgeoisie that is exercising dictatorship in our country. True enough, this is the period of the final struggle of dying imperialism—imperialism is "moribund capitalism." But just because it is dying, it is all the more dependent on colonies and semi-colonies for survival and will certainly not allow any colony or semi-colony to establish anything like a capitalist society under the dictatorship of its own bourgeoisie. Just because Japanese imperialism is bogged down in serious economic and political crises, just because it is dying, it must invade China and reduce her to a colony, thereby blocking the road to bourgeois dictatorship and national capitalism in China. . . .

[For the rest of Section VII, and through Sections VIII and IX, Mao spends six pages refuting alternate ideas for what China should do— from liberals, to extreme leftists, to die-hard conservatives.]

X. THE THREE PEOPLE'S PRINCIPLES, OLD AND NEW

The bourgeois die-hards have no understanding whatsoever of historical change; their knowledge is so poor that it is practically nonexistent. They do not know the difference either between communism and the Three People's Principles or between the new Three People's Principles and the old.

We Communists recognize "the Three People's Principles as the political basis for the Anti-Japanese United Front," we acknowledge that "the Three People's Principles being what China needs today, our Party is ready to fight for their complete realization," and we admit the basic agreement between the Communist minimum program and the political tenets of the Three People's Principles. But which kind of Three People's Principles? The Three People's Principles as reinterpreted by Mr. Sun Yat-Sen in the Manifesto of the First National Congress of the Guomindang, and no other. I wish the die-hard gentlemen would spare a moment from the work of "restricting communism," "corroding communism," and "combating communism," in which they are so gleefully engaged, to glance through this manifesto. In the manifesto Mr. Sun Yat-sen said: "Here is the true interpretation of the Guomindang's Three People's Principles." Hence these are the only genuine Three People's Principles and all others are spurious. The only "true interpretation" of the Three People's Principles is the one contained in the Manifesto of the First National Congress of the Guomindang, and all other interpretations are false. Presumably this is no Communist fabrication, for many Guomindang members and I myself personally witnessed the adoption of the manifesto.

The manifesto marks off the two epochs in the history of the Three People's Principles. Before it, they belonged to the old category; they were the Three People's Principles of the old bourgeois-democratic revolution in a semi-colony, the Three People's Principles of old democracy, the old Three People's Principles.

After it, they came within the new category; they became the Three People's Principles of the new bourgeois-democratic revolution in a semi-colony, the Three People's Principles of New Democracy, the new Three People's Principles. These and these alone are the revolutionary Three People's Principles of the new period.

The revolutionary Three People's Principles of the new period, the new or genuine Three People's Principles, embody the Three Great Policies of alliance with Russia, cooperation with the Communist Party, and assistance to the peasants and workers. Without each and every one of these Three Great Policies, the Three People's Principles become either false or incomplete in the new period.

In the first place, the revolutionary, new, or genuine Three People's Principles must include alliance with Russia. The present situation is perfectly clear. If there is no policy of uniting with Russia, if we do not unite with the land of socialism, there will inevitably be a policy of uniting with imperialism, we will inevitably unite with imperialism. Is it not evident that this is exactly what happened after 1927? During the first two years of the War of Resistance against Japan, because the Great Imperialist War had not yet broken out, the contradictions between Britain, the United States, and other countries and Japan could still be exploited. Since the outbreak of the Imperialist World War, these contradictions, although they have not entirely disappeared, have greatly diminished. If we were to make improper use of them, then England and the United States could demand that China participate in their struggle against the Soviet Union. If China then complied with their demand, she would immediately place herself on the side of the reactionary front of imperialism, thus putting an end to all national independence. Once the conflict between the socialist Soviet Union and imperialist Britain and the United States grows sharper, China will have to take her stand on one side or the other. This is an inevitable trend. Is it not possible to avoid leaning to either side? No, that is an illusion. The entire globe will be swept into one or the other of these two fronts, and henceforth "neutrality" will be merely a deceptive term. Especially is this true of China, which is fighting an imperialist power that has penetrated deep into her territory; her final victory is inconceivable without the assistance of the Soviet Union. If alliance with Russia is sacrificed for the sake of alliance with imperialism, the word "revolutionary" will have to be expunged from the Three People's Principles, which will then become reactionary. In the last analysis, there can be no "neutral" Three People's Principles; they can only be either revolutionary or counterrevolutionary. Would it not be more heroic to "fight against attacks from both sides" as Wang Jingwei once remarked, and to have the kind of Three People's Principles that serves this "fight"? Unfortunately, even its inventor Wang Jingwei himself has abandoned (or "folded up") this kind of Three People's Principles, for he has adopted the Three People's Principles of alliance with imperialism. If it is argued that there is a difference between Eastern and Western imperialism, and that, unlike Wang Jingwei who has allied himself with Eastern imperialism, you should ally yourself with some motherfucking Western imperialists to march eastward and attack, then would not such conduct be quite revolutionary? But whether you like it or not, the Western imperialists are determined to oppose the Soviet Union and communism, and if you ally yourself with them, they will ask you to march northward and attack, and your revolution will come to nothing. All these circumstances make it essential for the revolutionary, new, and genuine Three People's Principles to include alliance with Russia, and under no circumstances alliance with imperialism against Russia.

In the second place, the revolutionary, new, and genuine Three People's Principles must include cooperation with the Communist Party. Either you cooperate with the Communist Party or you oppose it. Opposition to communism is the policy of the Japanese imperialists and Wang Jingwei, and if that is what you want, very well, they will invite you to join their Anti-Communist Company. But wouldn't that look suspiciously like turning traitor? You may say, "I am not following Japan, but some other country." That is just ridiculous. No matter whom you follow, the moment you oppose the Communist Party you become a traitor, because you can no longer resist Japan. If you say, "I am going to oppose the Communist Party independently," that is arrant nonsense. How can the "heroes" in a colony or semi-colony tackle a counterrevolutionary job of this magnitude without depending on the strength of imperialism? For ten long years, virtually all the imperialist forces in the world were enlisted against the Communist Party, but in vain. How can you suddenly oppose it "independently"?

Some people outside the Border Region, we are told, are now saying: "Opposing the Communist Party is good, but you can never succeed in it." This remark, if it is not simply hearsay, is only half wrong, for what "good" is there in opposing the Communist Party? But the other half is true, you certainly can "never succeed in it." Basically, the reason lies not with the Communists but with the common people, who like the Communist Party and do not like "opposing" it. If you oppose the Communist Party at a juncture when our national enemy is penetrating deep into our territory, the people will be after your hide; they will certainly show you no mercy. This much is certain, whoever wants to oppose the Communist Party must be prepared to be ground to dust. If you are not keen on being ground to dust, you had certainly better drop this opposition. This is our sincere advice to all the anti-Communist "heroes." Thus it is as clear as can be that the Three People's Principles of today must include cooperation with the Communist Party, or otherwise those Principles will perish. It is a question of life and death for the Three People's Principles. Cooperating with the Communist Party, they will survive; opposing the Communist Party, they will perish. Can anyone prove the contrary?

In the third place, the revolutionary, new, and genuine Three People's Principles must include the policy of assisting the peasants and workers. Rejection of this policy, failure whole-heartedly to assist the peasants and workers or failure to carry out the behest in Mr. Sun Yat-sen's Testament to "arouse the masses of the people," amounts to preparing the way for the defeat of the revolution, and one's own defeat into the bargain. Stalin has said that "in essence, the question of colonies and semi-colonies is a peasant question."[16] This means that the Chinese revolution is essentially a peasant revolution and that the resistance to Japan now going on is essentially peasant resistance. Essentially, the politics of New Democracy means giving the peasants their rights. The new and genuine Three People's Principles are essentially the principles of a peasant revolution. Essentially, mass culture means raising the cultural level of the peasants. The anti-Japanese war is essentially a peasant war. We are now living in a time when the "doctrine of going up to the

[16]Translated in J. V. Stalin, "Concerning the National Question in Yugoslavia," in *Works*, 7:71–72.

mountains"[17] applies; everyone is on the top of the hills; meetings, work, classes, newspaper publication, the writing of books, theatrical performances—everything is done up in the hills, and all essentially for the sake of the peasants. And essentially it is the peasants who provide everything that sustains the resistance to Japan and keeps us going. By "essentially" we mean basically, not ignoring the other sections of the people, as Stalin himself has explained. As every schoolboy knows, 80 percent of China's population are peasants, more than 80 percent after the fall of the big cities. So the peasant problem becomes the basic problem of the Chinese revolution and the strength of the peasants is the main strength of the Chinese revolution. In the Chinese population the workers rank second to the peasants in number. There are several million industrial workers in China and several tens of millions of handicraft workers and agricultural laborers. China cannot live without them, because they are the producers in the industrial sector of the economy. And the revolution cannot succeed without them, because they are the leaders of the Chinese revolution and the most revolutionary class. In these circumstances, the revolutionary, new, and genuine Three People's Principles must include the policy of assisting the peasants and workers. Any other kind of Three People's Principles which lack this policy, do not give the peasants and workers wholehearted assistance, or do not carry out the behest to "arouse the masses of the people" will certainly perish.

Thus it is clear that there is no future for any Three People's Principles which depart from the Three Great Policies of alliance with Russia, cooperation with the Communist Party, and assistance to the peasants and workers. Every conscientious follower of the Three People's Principles must seriously consider this point.

The Three People's Principles comprising the Three Great Policies—in other words, the revolutionary, new, and genuine Three People's Principles—are the Three People's Principles of New Democracy, a development of the old Three People's Principles, a great contribution of Mr. Sun Yat-sen, and a product of the era in which the Chinese revolution has become part of the world socialist revolution. It is only the Three People's Principles which the Chinese Communist Party regards as "being what China needs today" and for whose "complete realization" it declares itself pledged "to fight." These are the only Three People's Principles which are in basic agree-

---

[17] *going up to the mountains:* forming base areas in remote places, such as Shaan-Gan-Ning

ment with the Communist Party's political program for the stage of democratic revolution, namely with its minimum program.

As for the old Three People's Principles, they were a product of the old period of the Chinese revolution. Russia was then an imperialist power, and naturally there could be no policy of alliance with her; there was then no Communist Party in existence in our country and naturally there could be no policy of cooperation with it; the movement of the workers and peasants had not yet revealed its full political significance and aroused people's attention, and naturally there could be no policy of alliance with them. Hence the Three People's Principles of the period before the reorganization of the Guomindang in the thirteenth year of the Republic [1924] belonged to the old category, and they became obsolete. The Guomindang could not have gone forward unless it had developed them into the new Three People's Principles. Mr. Sun Yat-sen in his wisdom saw this point, secured the help of Lenin and the Chinese Communist Party, and reinterpreted the Three People's Principles so as to endow them with new characteristics suited to the times. As a result, a united front was formed between the Three People's Principles and communism, Guomindang-Communist cooperation was established for the first time, the sympathy of the people of the whole country was won, and the First Great Revolution was launched.

The old Three People's Principles were revolutionary in the old period and reflected its historical features. But if the old stuff is repeated in the new period after the new Three People's Principles have been established, or alliance with Russia is opposed after the socialist state has been established, or cooperation with the Communist Party is opposed after the Communist Party has come into existence, or the policy of assisting the peasants and workers is opposed after they have awakened and demonstrated their political strength, then that is reactionary and shows ignorance of the times. The period of reaction after 1927 was the result of such ignorance. The old proverb says, "Whosoever understands the signs of the times is a great man." I hope the followers of the Three People's Principles today will bear this in mind.

Were the Three People's Principles to fall within the old category, then they would have nothing basically in common with the Communist minimum program, because they would belong to the past and be obsolete. Any sort of Three People's Principles that oppose Russia, the Communist Party, or the peasants and workers are definitely reactionary; they not only have absolutely nothing in common with the

Communist minimum program but are the enemy of communism, and there is no common ground at all. This, too, the followers of the Three People's Principles should carefully consider.

In any case, people with a conscience will never forsake the new Three People's Principles until the task of opposing imperialism and feudalism is basically accomplished. The only ones who do are people like Wang Jingwei. No matter how energetically they push their spurious Three People's Principles which oppose Russia, the Communist Party, and the peasants and workers, there will surely be no lack of people with a conscience and sense of justice who will continue to support Sun Yat-sen's genuine Three People's Principles. Many followers of the genuine Three People's Principles continued the struggle for the Chinese revolution even after the reaction of 1927, and the numbers will undoubtedly swell to tens upon tens of thousands now that the national enemy has penetrated deep into our territory. We Communists will always persevere in long-term cooperation with all the true followers of the Three People's Principles and, while rejecting the traitors and the sworn enemies of communism, will never forsake any of our friends.

XI. THE CULTURE OF NEW DEMOCRACY

In the foregoing we have explained the historical characteristics of Chinese politics in the new period and the question of the new democratic republic. We can now proceed to the question of culture.

A given culture is the ideological reflection of the politics and economics of a given society. There is in China an imperialist culture which is a reflection of imperialist rule, or partial rule, in the political and economic fields. This culture is fostered not only by the cultural organizations run directly by the imperialists in China but by a number of Chinese who have lost all sense of shame. Into this category falls all culture embodying a slave ideology. China also has a semi-feudal culture which reflects her semi-feudal politics and economy and whose exponents include all those who advocate the worship of Confucius, the study of the Confucian canon, the old ethical code, and the old ideas in opposition to the new culture and new ideas. Imperialist culture and semi-feudal culture are devoted brothers and have formed a reactionary cultural alliance against China's new culture. This kind of reactionary culture serves the imperialists and the feudal class and must be swept away. Unless it is swept away, no new culture of any kind can be built up. There is no construction without destruction, no flowing without damming, and no motion without rest; the two are locked in a life-and-death struggle.

As for the new culture, it is the ideological reflection of the new politics and the new economy which it sets out to serve.

As we have already stated in Section III, Chinese society has gradually changed in character since the emergence of a capitalist economy in China; it is no longer an entirely feudal but a semi-feudal society, although the feudal economy still predominates. Compared with the feudal economy, this capitalist economy is a new one. The political forces of the bourgeoisie, the petty bourgeoisie, and the proletariat are the new political forces which have emerged and grown simultaneously with this new capitalist economy. Various revolutionary parties, the Guomindang and the Communist Party being the most important among them, are the representatives of the awakened bourgeoisie, the petty bourgeoisie, and the proletariat. And the new culture reflects these new economic and political forces in the field of ideology and serves them. Without the capitalist economy, without the bourgeoisie, the petty bourgeoisie, and the proletariat, and without the political parties of these classes, the new ideology or new culture could not have emerged.

These new political, economic, and cultural forces are all revolutionary forces which are opposed to the old politics, the old economy, and the old culture. The old is composed of two parts, one being China's own semi-feudal politics, economy, and culture, and the other the politics, economy, and culture of imperialism, with the latter heading the alliance. Both are bad and should be completely destroyed. The struggle between the new and the old in Chinese society is a struggle between the new forces of the people (the various revolutionary classes) and the old forces of imperialism and the feudal class. It is a struggle between revolution and counterrevolution. This struggle has lasted a full hundred years starting from the Opium War, and nearly thirty years starting from the Revolution of 1911.

But as already indicated, revolutions too can be classified into old and new, and what is new in one historical period becomes old in another. The century of China's bourgeois-democratic revolution can be divided into two main stages, a first stage of eighty years and a second of twenty years. Each has its basic historical characteristics. China's bourgeois-democratic revolution in the first eighty years belongs to the old category, while in the last twenty years, owing to the change in the international and domestic political situation, it belongs to the new category. Old democracy is the characteristic of the first eighty years. New Democracy is the characteristic of the last twenty. This distinction holds good in culture as well as in politics.

How does it manifest itself in the field of culture? We shall explain this next.

XII. THE HISTORICAL CHARACTERISTICS OF CHINA'S CULTURAL REVOLUTION

On the cultural or ideological front, the two periods preceding and following the May Fourth Movement form two distinct historical periods.

Before the May Fourth Movement, the struggle on China's cultural front was one between the new culture of the bourgeoisie and the old culture of the feudal class. The struggles between the modern school system and the imperial examination system, between the new learning and the old learning, and between Western learning and Chinese learning, were all of this nature. The so-called modern schools or new learning or Western learning of that time concentrated mainly (we say mainly, because in part pernicious vestiges of Chinese feudalism still remained) on the bourgeois natural sciences and social sciences. In addition to the natural sciences at the time, the new schools before the May Fourth Movement were dominated by the ideology represented by Darwin's theory of evolution, Adam Smith's classical economics, Mill's formal logic, and French Enlightenment scholar Montesquieu's socialism introduced to China by Yan Fu.[18] At the time, this ideology played a revolutionary role in fighting the Chinese feudal ideology, and it served the bourgeois-democratic revolution of the old period. But because the Chinese bourgeoisie lacked strength and the world had already entered the era of imperialism, this bourgeois ideology was only able to last out a few rounds and was beaten back by the reactionary alliance of the enslaving ideology of foreign imperialism and the "back to the ancients" ideology of Chinese feudalism; as soon as this reactionary ideological alliance started a minor counteroffensive, the so-called new learning lowered its banners, muffled its drums, and beat a retreat, retaining its outer form but losing its soul. The old bourgeois-democratic culture became enervated and decayed in the era of imperialism, and its failure was inevitable.

[18]*Yan Fu:* the translator and interpreter of some of the most influential European thinkers at the turn of the twentieth century. Mao and others read Yan's translations of Charles Darwin, Herbert Spencer, John Stuart Mill, and others.

But since the May Fourth Movement things have been different. A brand-new cultural force came into being in China, that is, the Communist culture and ideology guided by the Chinese Communists, or the Communist world outlook and theory of social revolution. The May Fourth Movement occurred in 1919, and in 1921 came the founding of the Chinese Communist Party and the real beginning of China's labor movement—all in the wake of the First World War and the October Revolution, i.e., at a time when the national problem and the colonial movements of the world underwent a change, and the connection between the Chinese revolution and the world revolution became quite obvious. The new political force of the proletariat and the Communist Party mounted the Chinese political stage, and as a result, the new cultural force, in new uniform and with new weapons, mustering all possible allies and deploying its ranks in battle array, launched heroic attacks on imperialist culture and feudal culture. Although this vital force has not yet had the time to occupy the field of natural sciences and carry out a fight in it, in general allowing the bourgeois world outlook to dominate it temporarily, it has aroused a great revolution in the social science field, which provides the most important ideological weapons in the era of revolutions in the colonies and semi-colonies. This new force has made great strides in the domain of the social sciences, whether of philosophy, economics, political science, military science, history, literature, or art (including the theater, the cinema, music, sculpture, and painting). For the last twenty years, wherever this new cultural force has directed its attack, a great revolution has taken place both in ideological content and in form (for example, in the written language[19]). Its influence has been so great and its impact so powerful that it is invincible wherever it goes. The numbers it has rallied behind it have no parallel in Chinese history. Lu Xun[20] was the greatest and the most courageous standard-bearer of this new cultural force. The chief commander of China's cultural revolution, he was not only a great man of letters but a great thinker and revolutionary. Lu Xun was a man of unyielding integrity, free from all sycophancy or obsequiousness; this quality is invaluable among colonial and semi-colonial peoples. Representing the great

[19]Mao is referring to changes in written Chinese from the complex classical language to the vernacular, or *baihua*. This was a major cultural change, as it contributed to a rapid rise in popular literacy.

[20]*Lu Xun:* China's most famous man of letters in the 1920s and 1930s; considered China's foremost writer of the twentieth century. He was a biting social critic and came to support the Communist cause, although he never joined the CCP.

majority of the nation, Lu Xun breached and stormed the enemy citadel; on the cultural front he was the bravest and most correct, the firmest, the most loyal and the most ardent national hero, a hero without parallel in our history. The road he took was the very road of China's new national culture.

Prior to the May Fourth Movement, China's new culture was a culture of the old-democratic kind and part of the capitalist cultural revolution of the world bourgeoisie. Since the May Fourth Movement, it has become new-democratic and part of the socialist cultural revolution of the world proletariat.

Prior to the May Fourth Movement, China's new cultural movement, her cultural revolution, was led by the bourgeoisie, which still had a leading role to play. After the May Fourth Movement, the culture and ideology of this class became even more backward than its politics, and it was incapable of playing any leading role; at most, it could serve to a certain extent as an ally during revolutionary periods, while inevitably the responsibility for leading the alliance rested on proletarian culture and ideology. This is an undeniable fact.

The new-democratic culture is the anti-imperialist and anti-feudal culture of the broad masses; today it is the culture of the Anti-Japanese United Front. This culture can be led only by the culture and ideology of the proletariat, by the ideology of communism, and not by the culture and ideology of any other class. In a word, new-democratic culture is the proletarian-led, anti-imperialist, and anti-feudal culture of the broad masses.

XIII. THE FOUR PERIODS

A cultural revolution is the ideological reflection of the political and economic revolutions, and serves them. In China there is a united front in the cultural as in the political revolution.

The history of the united front in the cultural revolution during the last twenty years can be divided into four periods. The first covers the two years from 1919 to 1921, the second the six years from 1921 to 1927, the third the nine years from 1927 to 1936, and the fourth the three years from 1937 to the present.

The first period extended from the May Fourth Movement of 1919 to the founding of the Chinese Communist Party in 1921. The May Fourth Movement was its chief landmark.

The May Fourth Movement was an anti-imperialist as well as an anti-feudal movement. Its outstanding historical significance is to be seen in a feature which was absent from the Revolution of 1911, namely its thorough and uncompromising opposition to imperialism as well as to feudalism. The May Fourth Movement possessed this quality because capitalism had developed a step further in China and because new hopes had arisen for the liberation of the Chinese nation as China's revolutionary intellectual class saw the collapse of three great imperialist powers—Russia, Germany, and Austria-Hungary—and the weakening of two others, Britain and France, while the Russian proletariat had established a socialist state and the German, Hungarian, and Italian proletariat had risen in revolution. The May Fourth Movement came into being at the call of the world revolution of the time, of the Russian Revolution, and of Lenin. It was part of the world proletarian revolution of the time. Although at the time of the May Fourth Movement the Chinese Communist Party had not yet come into existence, there were already large numbers of intellectuals who approved of the Russian Revolution and had the rudiments of Communist ideology. In the beginning the May Fourth Movement was a revolutionary movement of the united front of three sections of people—Communist intellectuals, revolutionary petty-bourgeois intellectuals, and bourgeois intellectuals (the last forming the right wing at that time). Its weak point was that it was confined to the intellectuals, and the workers and peasants did not participate in it. But as soon as it developed into the June Third Movement,[21] not only the intellectuals but the mass of the proletariat, the petty bourgeoisie, and the bourgeoisie joined in, and it became a nationwide revolutionary movement. The cultural revolution ushered in by the May Fourth Movement was uncompromising in its opposition to feudal culture; there had never been such a great and thoroughgoing cultural revolution since the dawn of Chinese history. Raising aloft the two great banners of the day, "Down with the old ethics and up with the new!" and "Down with the old literature and up with the new!" the cultural revolution had great achievements to its credit. At that time it was not yet possible for this cultural movement to become widely diffused among the workers and peasants. The slogan of "Literature for the common people" was

---

[21]*June Third Movement:* an important extension of the demonstrations of May 4, 1919, in which students linked with workers to hold labor strikes in defiance of police repression

advanced, but in fact the "common people" then could only refer to the petty-bourgeois and bourgeois intellectuals in the cities, that is, the so-called urban intelligentsia. Both in ideology and in the matter of cadres, the May Fourth Movement paved the way for the founding of the Chinese Communist Party in 1921, for the May Thirtieth Movement of 1925,[22] and for the Northern Expedition. The bourgeois intellectuals, who constituted the right wing of the May Fourth Movement, mostly compromised with the enemy in the second period and went over to the side of reaction.

In the second period, whose landmarks were the founding of the Chinese Communist Party, the May Thirtieth Movement, and the Northern Expedition, the united front of the three classes formed during the May Fourth Movement was continued and expanded. This united front also took form politically, this being the first instance of Guomindang-Communist cooperation. Mr. Sun Yat-sen was a great man not only because he led the great Revolution of 1911 (although it was only a democratic revolution of the old period), but also because, "adapting himself to the trends of the world and meeting the needs of the masses," he had the capacity to bring forward the revolutionary Three Great Policies of alliance with Russia, cooperation with the Communist Party, and assistance to the peasants and workers, give new meaning to the Three People's Principles, and thus institute the new Three People's Principles with their Three Great Policies. Previously, the Three People's Principles had exerted little influence on the educational and academic world or with youth, because they had not raised the issues of opposition to imperialism or to the feudal social system and feudal culture. They were the old Three People's Principles which people regarded as the time-serving banner of a group of men bent on seizing power, in other words, on securing official positions, a banner used purely for political maneuvering. Then came the new Three People's Principles with their Three Great Policies. The cooperation between the Guomindang and the Communist Party and the joint efforts of the revolutionary members of the two parties spread the new Three People's Principles all over China, extending to a section of the educational and academic world and the mass of student youth. This was entirely because the original Three People's Principles had developed into the anti-imperialist, anti-feudal,

[22] *May Thirtieth Movement of 1925:* started with a violent confrontation between protesters and British soldiers in Shanghai and touched off nationwide demonstrations and boycotts against foreign goods in China

and new-democratic Three People's Principles with their Three Great Policies. Without this development it would have been impossible to disseminate the ideas of the Three People's Principles.

During this period, the revolutionary Three People's Principles became the political basis of the united front of the Guomindang and the Communist Party and of all the revolutionary classes, and since "communism is the good friend of the Three People's Principles," a united front was formed between the two of them. In terms of social classes, it was a united front of three classes of the proletariat, the petty bourgeoisie, and the bourgeoisie. Using the Guomindang's *Republican Daily News* of Shanghai and other newspapers in various localities as their bases of operations, the two parties jointly advocated anti-imperialism, jointly combated feudal education based upon the worship of Confucius and upon the study of the Confucian canon, and jointly opposed feudal literature and the classical language and promoted the new literature and the vernacular style of writing with an anti-imperialist and anti-feudal content. During the wars in Guangdong and during the Northern Expedition, they reformed China's armed forces by the inculcation of anti-imperialist and anti-feudal ideas. The slogans, "Down with the corrupt officials" and "Down with the local bullies and bad gentry," were first raised among the peasant millions, and great peasant revolutionary struggles were aroused. Thanks to all this and to the assistance of the Soviet Union, the Northern Expedition was victorious. But no sooner did the bourgeoisie climb to power with the help of workers and peasants than it put an end to this great revolution, thus creating an entirely new political situation.

The third period was the new revolutionary period of 1927–36. As a result of the changes which had taken place within the revolutionary camp at the end of the previous period, with the bourgeoisie going over to the counterrevolutionary camp of the imperialist and feudal forces, only two of the three classes formerly within the revolutionary camp remained, namely, the proletariat and the petty bourgeoisie (including the peasantry, the revolutionary intellectuals, and other sections of the petty bourgeoisie). Thus the Chinese revolution inevitably entered a new period in which the Chinese Communist Party alone exercised the leadership. This period was one of reactionary campaigns of "encirclement and suppression," on the one hand, and of the deepening of the revolution, on the other. There were two kinds of reactionary campaigns of "encirclement and suppression," the military and the cultural. The deepening of the revolution was of two kinds; both the agrarian and the cultural revolutions were deepened. At the instigation of the imperialists, the reactionary forces of the whole country and of the whole world were mobilized for both kinds of campaigns of "encirclement and suppression," which lasted no less than ten years and were unparalleled in their ruthlessness; hundreds of thousands of Communists and young students were slaughtered and millions of workers and peasants suffered cruel persecution. The people responsible for all this apparently had no doubt that communism and the Communist Party could be "exterminated once and for all." However, the outcome was different; both kinds of "encirclement and suppression" campaigns failed miserably. The military campaign resulted in the northern march of the Red Army to resist the Japanese,[23] and the cultural campaign resulted in the outbreak of the December Ninth movement of the revolutionary youth in 1935. And the common result of both was the awakening of the people of the whole country. These were three positive results. But the negative result was the attack by a powerful enemy; this is the key reason why the people of the whole country to this day bitterly detest the anti-communism of those ten years. The most amazing thing of all was that the Guomindang's cultural "encirclement and suppression" campaign failed completely in the Guomindang areas as well, although the Communist Party was in an utterly defenseless position in all the cultural and educational institutions there. Why did this happen? Does it not give food for prolonged and deep thought? It was in the very midst of such campaigns of "encirclement and suppression" that Lu Xun, who believed in communism, became the giant of China's cultural revolution.

In the struggles of this period, the revolutionary side firmly upheld the people's anti-imperialist and anti-feudal New Democracy and their new Three People's Principles, while the counterrevolutionary side, under the direction of imperialism, imposed the despotic regime of the coalition of the landlord class. That despotic regime butchered Mr. Sun Yat-sen's Three Great Policies and his new Three People's Principles both politically and culturally, with catastrophic consequences to the Chinese nation.

The fourth period is that of the present anti-Japanese war. Pursuing its zigzag course, the Chinese revolution has again arrived at a united front of the three classes. But this time the scope is much broader.

[23] Mao is referring to the Long March of 1934–35, which ultimately brought the main CCP forces to Yan'an.

Among the upper classes it includes all the rulers. Among the middle classes, it includes the petty bourgeoisie in its totality. Among the lower classes, it includes the entire proletariat. All classes and strata of the country have become allies, and are resolutely resisting Japanese imperialism. The first stage of this period lasted until the fall of Wuhan. During that stage, there was a lively atmosphere in the country in every field; politically there was a trend toward democracy and culturally there was widespread activity. With the fall of Wuhan [in October 1938] the second stage began, during which the political situation has undergone many changes, with one section of the big bourgeoisie capitulating to the enemy and another desiring an end to the War of Resistance. In the cultural movement, this situation has been reflected in the reactionary activities of Ye Qing,[24] Zhang Junmai, and others, and in the suppression of freedom of speech and of the press.

To overcome this crisis, a firm struggle is necessary against all ideas opposed to resistance, unity, and progress, and unless these reactionary ideas are crushed, there will be no hope of victory. How will this struggle turn out? This is the big question in the minds of the people of the whole country. Judging by the domestic and international situation, the Chinese people are bound to win, however numerous the obstacles on the path of resistance. If we consider Chinese history as a whole, the progress achieved during the twenty years since the May Fourth Movement not only surpasses that of the preceding eighty years, it truly surpasses that achieved of several millennia. Can we not visualize what further progress China will make in another twenty years? The unbridled violence of all the forces of darkness, whether domestic or foreign, has brought disaster to our nation; but this very violence indicates that while the forces of darkness still have some strength left, they are already in their death throes, and that the people are gradually approaching victory. This is true in the East and in the entire world. . . .

[In Section XIV Mao reviews in about two pages what he sees as wrong ideas about culture and how Communists and their supporters should struggle against these errors by spreading Marxism-Leninism.]

---

[24] *Ye Qing:* a former leftist who used Marxist theory to attack CCP policies

XV. A NATIONAL, SCIENTIFIC, AND MASS CULTURE

New-democratic culture is national. It opposes imperialist oppression and upholds the dignity and independence of the Chinese nation. It belongs to this nation of ours, and bears our own national characteristics. It links up with the socialist and new-democratic cultures of other nations, and establishes with them the relations whereby they can absorb something from each other and help each other to develop, mutually forming a part of a new world culture. But it can absolutely not link up with any reactionary imperialist culture of whatever nation, for our culture is a revolutionary national culture. China must assimilate on a large scale the progressive culture of foreign countries, as an ingredient for enriching her own culture. Not enough of this was done in the past. We should assimilate whatever is useful to us today not only from the present-day socialist and new-democratic cultures, but also from the older cultures of foreign countries, for example, from the culture of the various capitalist countries in the Age of Enlightenment. However, we absolutely cannot gulp down any of this foreign material uncritically, but must treat it as we do our food—first chewing it in the mouth, then subjecting it to the working of the stomach and intestines with their juices and secretions, and separating it into essence to be absorbed and waste matter to be discarded—before it can nourish us. So-called "wholesale Westernization" is wrong. China has suffered a great deal in the past from the formalist absorption of foreign things. Similarly, in applying Marxism to China, Chinese Communists must fully and properly integrate the universal truth of Marxism with the concrete practice of the Chinese revolution, or in other words, the universal truth of Marxism must have a national form if it is to be useful, and in no circumstances can it be applied subjectively as a mere formula. Marxists who make a fetish of formulas are simply playing with Marxism and the Chinese revolution, and there is no room for them in the ranks of the Chinese revolution. Chinese culture should have its own form, its own national form. National in form and new-democratic in content—such is our new culture today.

New-democratic culture is scientific. Opposed as it is to all feudal and superstitious ideas, it stands for "seek truth from facts," for objective truth, and for the unity of theory and practice. On this point, the possibility exists of a united front against imperialism, feudalism, and superstition between the scientific thought of the Chinese proletariat and those Chinese bourgeois materialists and natural scientists who are progressive, but in no case is there a possibility of a united front with any reactionary idealism. In the field of political action Communists may form an anti-imperialist united front with some idealists and even religious people, but we can never approve of their idealism or religious doctrines. A splendid ancient culture was created during the long period of China's feudal society. Therefore, to clarify the process of development of this ancient culture, to discard its feudal dross and assimilate its democratic essence, is a necessary condition for developing our new national culture and increasing our national self-confidence, but we absolutely cannot swallow anything and everything uncritically. We must separate the fine, old popular culture which had a more or less democratic and revolutionary character from all the rotten things of the old feudal ruling class. China's present new politics and new economy have developed out of her old politics and old economy, and her present new culture, too, has developed out of her old culture. Consequently, we must respect our own history and can absolutely not mutilate history. Respect for history means, however, giving it its proper place as a science, respecting its dialectical development, and not eulogizing the past at the expense of the present or praising every drop of feudal poison. As far as the masses and the young students are concerned, the essential thing is to guide them to look forward and not backward.

New-democratic culture belongs to the broad masses and is therefore democratic. It should serve the toiling masses of workers and peasants who make up more than 90 percent of the nation's population and should gradually become their very own. There is a difference of degree, as well as a close link, between the knowledge imparted to the revolutionary cadres and the knowledge imparted to the revolutionary masses, between the raising of cultural standards and popularization. Revolutionary culture is a powerful revolutionary weapon for the broad masses of the people. It prepares the ground ideologically before the revolution comes and is an important, indeed essential, fighting front in the general revolutionary front during the revolution. People engaged in revolutionary cultural work are the commanders at various levels on this cultural front. "Without revolutionary theory there can be no revolutionary movement"; one can thus see how important the cultural movement is for the practical revolutionary movement. Both the cultural and practical movements must be of the masses. Therefore all progressive cultural workers in the anti-Japanese war must have their own cultural battalions, that is, the broad masses. A cultural worker or a cultural ideology detached from the popular masses is a "shadow" commander without an army, whose firepower cannot bring the enemy down. To attain this objective, written Chinese must be reformed, given the requisite conditions, and our spoken language brought closer to that of the people, for the people, it must be stressed, are the inexhaustible source of our revolutionary culture.

A national, scientific, and mass culture—such is the anti-imperialist and anti-feudal culture of the people, the culture of New Democracy and the new Three People's Principles, the new culture of the Chinese nation.

Combine the politics, the economy, and the culture of New Democracy, and you have the new-democratic republic, the Republic of China both in name and in reality, the new China we want to create.

Behold, New China is within sight. Let us all hail her!
Her masts have already risen above the horizon. Let us all cheer in
  welcome!
Raise both your hands. New China is ours!

3

Talks at the Yan'an Conference
on Literature and Art

1942

As Mao consolidated his power over the CCP in Yan'an in the early 1940s, he was unexpectedly confronted by biting criticism of the party's failure to live up to its own egalitarian goals. These criticisms came from the CCP's left-wing intellectuals, hundreds of whom had come to Yan'an after the outbreak of war with Japan in 1937. In May 1942, Mao paused the Rectification Movement—his broader campaign to eliminate his senior party rivals and mobilize the party rank and file to the ideals of his "On New Democracy" (see Document 2)—to discipline his way-

"Zai Yan'an wenyi zuotanhui shangde jianghua," Jiefang ribao, 19 Oct. 1943. Translation from Bonnie S. McDougall, Mao Zedong's "Talks at the Yan'an Conference on Literature and Art": A Translation of the 1943 Text with Commentary (Ann Arbor: Michigan Studies in Chinese Studies, 1980), 57–58, 60–61, 69–70, 75. McDougall carefully notes the changes made for the official 1951 Selected Works edition in her endnotes, which are not retained here.

ward left-wing critics. The result was the Yan'an Conference on Literature and Art.

The brief extracts here from Mao's talks at the conference focus on two fundamental points: First, the importance of the social role of art (in service of the revolution) and the need for art to subordinate itself to politics. Mao's view of art is not romantic. He sees "cultural workers" as just another brigade in the revolutionary army. Yet he acknowledges that they are a necessary "screw" in the revolutionary machine. Second, Mao outlines what artists need to do to "serve the people." They need to "become one with the masses" by the physical act of living, working, and fighting against the Japanese with the poor farmers of China. Mao underlines this with a most astonishing autobiographical story of his "conversion" from bourgeois pride to solidarity with the working poor. It is a noble image of reaching out beyond one's social class, but it is chilling when we remember that Mao fully expected everyone to have the identical experience.

The translation here comes from the original published version of 1943 in Yan'an's party newspaper, Liberation Daily. *Bonnie McDougall's careful translation retains the feel of Mao's 1940s writings and his earthy style much better than the overly polished and sanitized 1951* Selected Works *version.*

INTRODUCTION

(2 May 1942)

Comrades! I have invited you to this conference today for the purpose of exchanging opinions with you on the correct relationship between work in literature and art and revolutionary work in general, to obtain the correct development of revolutionary literature and art and better assistance from them in our other revolutionary work, so that we may overthrow our national enemy and accomplish our task of national liberation.

There are a number of different fronts in our struggle for the national liberation of China, civil and military, or, we might say, there is a cultural as well as an armed front. Victory over the enemy depends primarily on armies with guns in their hands, but this kind of army alone is not enough. We still need a cultural army, since this kind of army is indispensable in achieving unity among ourselves and winning victory over the enemy. Since May Fourth, when this cultural

army took shape in China, it has aided the Chinese revolution by
gradually limiting the sphere of China's feudal culture and the slavish
culture that serves imperialist aggression, and weakening their
strength, so that now reactionaries are reduced to resisting new culture by "meeting quality with quantity": reactionaries aren't short of
money, and with some effort they can turn out a lot even if they can't
come up with anything worthwhile. Literature and art have formed an
important and successful part of the cultural front since May
Fourth. . . . Our meeting today is to ensure that literature and art
become a component part of the whole revolutionary machinery, so
they can act as a powerful weapon in uniting and educating the people
while attacking and annihilating the enemy, and help the people
achieve solidarity in their struggle against the enemy. What are the
problems which must be solved in order to achieve this purpose?
They are questions relating to our position, attitude, audience, work,
and study. . . .

Since the audience for literature and art consists of workers, peasants, soldiers, and their cadres, the question then arises of how to get
to understand and know these people properly. To do this, we must
carry out a great deal of work in Party and government organs, in villages and factories, in the Eighth Route Army and the New Fourth
Army,[1] getting to understand all sorts of situations and all sorts of
people and making ourselves thoroughly familiar with them. Our
workers in literature and art must carry out their own work in literature and art, but the task of understanding people and getting to know
them properly has the highest priority. How have our workers in literature and art performed in this respect until now? I would say that
until now they have been heroes without a battlefield, remote and
uncomprehending. What do I mean by remote? Remote from people.
Workers in literature and art are unfamiliar with the people they write
about and with the people who read their work, or else have actually
become estranged from them. Our workers in literature and art are
not familiar with workers, peasants, soldiers, or even their cadres.
What do I mean by uncomprehending? Not comprehending their language. Yours is the language of intellectuals, theirs is the language of
the popular masses. I have mentioned before that many comrades like

---

[1]The Eighth Route Army and the New Fourth Army were the two major military
forces of the CCP after the start of the Second United Front in 1937. The New Fourth
Army was all but destroyed by GMD forces in 1941, but the Eighth Route Army survived and grew to become the People's Liberation Army, which conquered all of China
by 1949.

to talk about "popularization," but what does popularization mean? It means that the thoughts and emotions of our workers in literature and art should become one with the thoughts and emotions of the great masses of workers, peasants, and soldiers. And to get this unity, we should start by studying the language of the masses. If we don't even understand the masses' language, how can we talk about creating literature and art? "Heroes without a battlefield" refers to the fact that all your fine principles are not appreciated by the masses. The more you parade your qualifications before the masses, the more you act like "heroes," and the harder you try to sell your principles to them, the more the masses will resist buying. If you want the masses to understand you, if you want to become one with the masses, you must make a firm decision to undergo a long and possibly painful process of trial and hardship. At this point let me relate my own experience in how feelings are transformed. I started off as a student at school, and at school I acquired student habits, so that I felt ashamed to do any manual labor such as carry my own bags in front of all those students who were incapable of carrying anything for themselves. I felt that intellectuals were the only clean people in the world, and that workers, peasants, and soldiers were in general rather dirty. I could wear clothes borrowed from an intellectual, because I considered them clean, but I would not wear workers', peasants', or soldiers' clothes, because I thought they were dirty. When I joined the revolution and lived among workers, peasants, and soldiers, I gradually became familiar with them and they got to know me in return. Then and only then the bourgeois and petty bourgeois feelings taught to me in bourgeois schools began to undergo a fundamental change. Comparing intellectuals who have not yet reformed with workers, peasants, and soldiers, I came to feel that intellectuals are not only spiritually unclean in many respects but even physically unclean, while the cleanest people are workers and peasants; their hands may be dirty and their feet soiled with cow dung, but they are still cleaner than the big and petty bourgeoisie. This is what I call a transformation in feelings, changing over from one class to another. If our workers in literature and art who come from the intelligentsia want their work to be welcomed by the masses, they must see to it that their thoughts and feelings undergo transformation and reform. Otherwise, nothing they do will turn out well or be effective. . . .

CONCLUSION

(23 May 1942)

... Revolutionary Chinese writers and artists, the kind from whom we expect great things, must go among the masses; they must go among the masses of workers, peasants, and soldiers and into the heat of battle for a long time to come, without reservation, devoting body and soul to the task ahead; they must go to the sole, the broadest, and the richest source, to observe, experience, study, and analyze all the different kinds of people, all the classes and all the masses, all the vivid patterns of life and struggle, and all literature and art in their natural form, before they are ready for the stage of processing or creating, where you integrate raw materials with production, the stage of study with the stage of creation. Otherwise, there won't be anything for you to work on, since without raw materials or semiprocessed goods you have nothing to process and will inevitably end up as the kind of useless writer or artist that Lu Xun in his will earnestly instructed his son never to become.

... In the world today, all culture or literature and art belongs to a definite class and party, and has a definite political line. Art for art's sake, art that stands above class and party, and fellow-travelling or politically independent art do not exist in reality. In a society composed of classes and parties, art obeys both class and party and it must naturally obey the political demands of its class and party, and the revolutionary task of a given revolutionary age; any deviation is a deviation from the masses' basic need. Proletarian literature and art are a part of the whole proletarian revolutionary cause; as Lenin said, they are "a screw in the whole machine,"[2] and therefore, the party's work in literature and art occupies a definite, assigned position within the party's revolutionary work as a whole. Opposition to this assignment must lead to dualism or pluralism, and in essence resembles Trotsky's "Politics—Marxist; art—bourgeois."[3] We do not support excessive emphasis on the importance of literature and art, nor do we

[2]Lenin, in his 1902 goals for press and propaganda, used this mechanical metaphor, with soldiers, politicians, and publicists serving as coordinated parts of the revolutionary machine. Mao goes farther than Lenin by extending this machine to include creative literature.

[3]Leon Trotsky, the hero of the Russian Revolution who lost out to Stalin in factional fighting in the late 1920s, was known for his liberal views on fine arts and literature. Mao characterizes them as "politics—Marxist; art—bourgeois" to emphasize the lack of control over art and literature favored by Trotsky. By the 1940s, anything associated with Trotsky was bad news inside the Communist movement.

support their underestimation. Literature and art are subordinate to politics, and yet in turn exert enormous influence on it. Revolutionary literature and art are a part of the whole work of revolution; they are a screw, which of course doesn't compare with other parts in importance, urgency, or priority, but which is nevertheless indispensable in the whole machinery, an indispensable part of revolutionary work as a whole. If literature and art did not exist in even the broadest and most general sense, the revolution could not advance or win victory; it would be incorrect not to acknowledge this. Furthermore, when we speak of literature and art obeying politics, politics refers to class and mass politics and not to the small number of people known as politicians. Politics, both revolutionary and counterrevolutionary alike, concerns the struggle between classes and not the behavior of a small number of people. Ideological warfare and literary and artistic warfare, especially if these wars are revolutionary, are necessarily subservient to political warfare, because class and mass needs can only be expressed in a concentrated form through politics.

4

Resolution of the Central Committee of the Chinese Communist Party on Methods of Leadership

June 1, 1943

This resolution of the CCP's politburo of the Central Committee was passed in Yan'an on June 1, 1943. It is attributed to Mao Zedong and is included in volume 3 of his Selected Works *as "Some Questions Concerning Methods of Leadership." We have every reason to believe that he did write it or that it captures what he was telling his comrades. The resolution sums up the organizational lessons of the 1942–44 Rectification Movement in Yan'an and outlines in some detail how the party should organize mass mobilization. It is a blueprint of how to run the revolution at the local level, thus answering the challenge Mao set forth in his 1927*

"Zhonggong zhongyang guanyu lingdao fangfade jueding," in *Zhengfeng wenxian* (Rectification Documents) (Yan'an, 1944). Translation from Boyd Compton, ed., *Mao's China: Party Reform Documents, 1942–44* (Seattle: University of Washington Press, 1952), 176–83.

"Report on the Peasant Movement in Hunan" (see Document 1). It was extremely effective in the 1940s as the CCP extended its sway into new regions of China.

The lessons of coordinated but flexible organizing outlined in the resolution have been applied to social movements elsewhere, from the Vietcong in Vietnam to Che Guevara in Latin America to Marxist insurgents in Nepal. The key points are (1) a version of "think globally, act locally," but with a strong Leninist chain of command; (2) a hardheaded assessment of the "masses" one wants to mobilize (usually, 10 percent activists, 80 percent average, 10 percent backward); (3) a focus on nurturing that activist 10 percent to get the movement going; and (4) the importance of coordinated propaganda to guide leadership and motivate the rank and file. The philosophical method of this approach to changing society requires "theory-practice-theory," in which an ideology (Marxism) is tested by actual efforts to do something and then modified on the basis of the practical results of one's efforts. (Nick Knight provides a helpful chart of this process in Document 13.)

The most famous phrase from this resolution is the populist credo of Maoism: "Correct leadership must come from the masses and go to the masses." When the CCP followed this heartfelt populist method by taking the time to research local conditions and talk in advance to local people, it was extremely successful. When the increasingly powerful CCP apparatus bypassed the laborious "people's democratic methods" outlined in the resolution, it made mistakes, culminating in Mao's colossal errors in the Great Leap Forward (see Documents 8 and 14).

(Passed by the Political Bureau of the Central Committee, June 1, 1943)

1. Two methods must be adopted in accomplishing any task: the first is to combine the general and the particular, and the second is to unite leadership with the masses.

2. If any work or mission lacks a general, universal slogan, the broad masses cannot be moved to action, but if there is nothing more than a general slogan and the leaders do not make a concrete, direct, and thorough application of it with those from a particular unit who have been rallied around the slogan, [if the leaders] fail to break through at some point and gain experience, or fail to use acquired experience in later guiding other units, there is then no way for the leaders to test the correctness of the general slogan and there is no

way for them to carry out its contents; there is then the danger that the general slogan will have no effect.

For example, in the general reform of 1942, all those who met with success had adopted the method of combining a general slogan with particular guidance. All those who were unsuccessful had failed to adopt this method. In the reform movement of 1943, all Central Committee bureaus and subbureaus, cultural committees, and district and local Party committees must not only present general slogans (the reform plan for the entire year), but also acquire experience by selecting two or three units from their own organization, nearby organizations, schools, and military groups (the number need not be large), studying them thoroughly, gaining a detailed understanding of the process whereby reform and study have developed within these units, gaining a detailed understanding of the characteristics of the history, experience, thought, etc., the diligence in study and the quality of work of certain specific typical cadres from among these units (the number need not be large), and in addition personally guiding the leaders of these units to come to concrete solutions of the actual problems facing the units. The same should be done with a number of units within these organs, schools, or military groups, and the leaders of these organs, schools, or military groups should also utilize the method described above. This is also the method to be used by leaders to combine guidance and study. Any leader who, in his studies, fails to observe specific units and specific individuals at the lower levels or fails to refer to concrete cases, will never be able to give the units general guidance. This method must be universally promoted so that it can be mastered and applied by leaders and cadres at all levels.

3. The experience of the reform movement of 1942 has also proved that in the process of reform, the reform of each concrete unit must produce a leading nucleus of minority activists who are the core of the administrative leadership of that unit and it must also bring this leading nucleus into close union with the broad masses engaged in study; in this way only can reform fulfill its mission. If there is only a positive spirit on the part of the leading nuclei, it [the reform] becomes an empty flurry of activity on the part of a minority; yet if there is only a positive spirit on the part of the broad masses, with no powerful leading nucleus to organize the positive spirit of the masses properly, the masses' spirit then cannot endure, nor can it move in a correct direction or be elevated to a high standard.

Wherever there are masses, there are in all probability three groups: those who are comparatively active, those who are average, and those who are backward. In comparing the three groups, the two extremes are in all probability small, while the middle group is large. As a result, leaders must be skillful at consolidating the minority activists to act as a leading nucleus, and must rely on this nucleus to elevate the middle group and capture the backward elements. A truly consolidated, uniform, and united nucleus for the leadership of the masses must materialize gradually from the mass struggle (for example, in reform and study); it cannot materialize apart from the mass struggle.

In the process of any great struggle, the leading nucleus in the initial, intermediate, and final stages should not be, and cannot be, entirely the same; activists (heroes) in the struggle must be constantly recruited to replace those elements which were originally part of the nucleus, but which have been found wanting on closer inspection, or have degenerated. A fundamental reason that the work in many areas and many organs has not progressed has been the lack of just such a constantly healthy leading nucleus, which is unified and connected with the masses.

If a school of one hundred persons does not have among its teachers, experts, and students a leading nucleus of a few or a few dozen individuals, which is formed naturally (and not assembled by compulsion) by those who are comparatively the most active, orthodox, and intelligent, the school will then be difficult to manage. In all organs, schools, or units of any size, we should start with an application of Stalin's comments on the establishment of leadership, to be found in the ninth section of his discussion of the "Bolshevization of the Party." The standards for this leading nucleus should be the four points raised by Dimitrov[1] in his discussion of cadre policy (unlimited loyalty, relation to the masses, ability to do independent work, observation of discipline). Whether the mission is concerned with war, production, or education (including reform), or whether the work is reform and study, supervision of work, the investigation of cadres, or any other task, we must adopt not only the method of combining a general slogan with particular guidance, but also the method of combining the leading nuclei with the broad masses.

4. In all our Party's actual work, correct leadership must come from the masses and go to the masses. This means taking the views of the

[1]*Dimitrov:* Georgi Dimitrov, the leader of the Comintern, the Communist International that ran Russia's program for international Communist revolution from 1919 to 1943. Dimitrov was the key leader in the 1930s, and his writings were used by Mao and the CCP in Yan'an.

masses (unintegrated, unrelated views) and subjecting them to concentration (they are transformed through research into concentrated systematized views), then going to the masses with propaganda and explanation in order to transform the views of the masses, and seeing that these [views] are maintained by the masses and carried over into their activities. It also means an examination of mass activities to ascertain the correctness of these views. Then again, there is concentration from the masses and maintenance among the masses. Thus the process is repeated indefinitely, each time more correctly, vitally, and fruitfully. This is the epistemology and methodology of Marxism-Leninism.

5. The idea that correct relations should be created between the leading nucleus and the broad masses in organization, in the struggle, and in action; the idea that correct guiding views can result only from the process of concentrating from the masses and maintaining among the masses; and the idea of combining a general slogan with particular guidance when the views of the leadership are being carried out ... these ideas must be universally propagated in the current reform movement, so that mistaken views existing among the cadres on these questions can be corrected. Many comrades neglect or are unskillful in the task of consolidating activists and organizing a leading nucleus and neglect or are unskillful in the task of closely uniting the leading nucleus and the broad masses. As a consequence, their leadership becomes a bureaucratic leadership separated from the masses. Many comrades neglect or are unskillful at summarizing the experience of the mass struggle and delight in showing off by expressing many subjectivistic views; as a consequence, their own views become impractical nonsense. Many comrades are satisfied with the general slogan for their task, and either neglect or are unskillful at providing close, direct, specific, and concrete leadership after the slogan has been adopted. As a consequence, the slogan goes no further than talk, a piece of paper, or a meeting, and becomes bureaucratic leadership. The current reform movement must correct these defects. In reform of study, the supervision of work, and the investigation of cadres, [we must] learn the methods of uniting leadership and the masses and combining the general and the particular; we must then adopt these methods in all our work.

6. Correct guiding views are those which are concentrated from the masses then maintained among the masses; this is a fundamental method. In concentrating and maintaining, the method of combining a general slogan and particular guidance must be adopted. This is an

integral part of the foregoing method. From many specific [experi-
ences] of leadership, a general view is formulated (a general slogan):
this general view is then tested in particular units (you must not only
do this yourself, but must also ask others to do the same); then new
experiences can be concentrated (summarized experiences) and
established as a new guide for the general leadership of the masses.
Comrades should do this in the current reform movement and should
do the same in any work they undertake. Comparatively good leadership results from comparative skill in this method.

7. In any type of work (military work, production, education, reform
and study, supervision, the investigation of cadres or propaganda,
organizational work, counterespionage, etc.), higher-level guiding
organs should work through persons in responsible positions in lower-
level organs connected with that work, see that they assume responsibilities, bring about a division of labor, and at the same time achieve a
unified objective (centralization). It is not enough for the individual
departments merely to contact individual lower-level departments
(for example, the higher-level Organizational Bureau contacts only
the lower-level Organizational Bureau, the higher-level Propaganda
Bureau contacts only the lower-level Propaganda Bureau; the higher-
level Counterespionage Bureau contacts only the lower-level Counter-
espionge Bureau) so that those primarily responsible in lower-level
organs (for example, secretaries, chairmen, department heads, and
school principals, etc.) do not understand or accept responsibility;
they should see that both those with primary and secondary responsibilities understand and accept responsibility. . . .

8. According to the concrete historical and environmental conditions of each district, leaders should plan and control the general
program and make correct decisions on central tasks and the order of
work during a given period, sticking rigidly to this order until definite
results are achieved. This is one of the arts of leadership. It is also a
problem which must be closely considered in determining methods of
leadership, while applying the principles of uniting leadership with the
masses and combining the general and the particular.

9. Systematic attention has not been given here to questions of
detail concerning methods of leadership. On the basis of the principles and directives presented in this resolution, comrades in all
areas should reflect carefully and develop their own creative abilities.
The more bitter the struggle becomes, the more necessary is the

demand for close union between men of the Communist Party and the broad masses, the more necessary to Communist Party members is the close union between general slogans and particular guidance, and [the more necessary is] the thorough disruption of subjectivistic and bureaucratic methods of leadership. All leading comrades of our Party must forthwith adopt scientific methods of leadership and oppose them to subjective and bureaucratic methods of leadership, overcoming the latter with the former. Subjectivists and bureaucratists who do not understand the principles of uniting leadership with the masses and combining the general and the particular, greatly hamper the development of Party work. We must therefore oppose subjectivistic and bureaucratic methods of leadership, and universally and profoundly promote methods of leadership which are scientific.

5

Snow

1945

This is Mao's most famous poem, written in the Yan'an area of northwest China near the Yellow River, a dry and rugged region. Here Mao has adopted the voice of classical Chinese poetry, which is short and lyrical. The first stanza should be read like a Chinese painting—dabs of ink outlining the mental picture of Mao viewing this grandiose landscape. The second stanza reveals Mao's feelings. He stands where great emperors of the past have stood, both literally by the Yellow River and figuratively on the verge of taking power. His conclusion is resolutely revolutionary: "For men of vision/We must seek among the present generation." Written after his consolidation of power inside the CCP, as the Japanese war effort was flagging and the CCP "base areas" were growing, this poem conveys his confidence and elation.

Xue, in *Mao Zedong shici xuan* (Beijing: Renmin wenxue chubanshe, 1986), 61–62. Translated by Michael Bullock and Jerome Ch'en in Jerome Ch'en, *Mao and the Chinese Revolution* (London: Oxford University Press, 1965), 340–41. Chinese editions date this poem to 1936, but I follow Bullock and Ch'en's dating of 1945.

The northern scene:
A thousand leagues locked in ice,
A myriad leagues of fluttering snow.
On either side of the Great Wall
Only one vastness to be seen.
Up and down this broad river
Torrents flatten and stiffen.
The mountains* are dancing silver serpents
And hills, like waxen elephants, plod on the plain,
Challenging heaven with their heights.
A sunny day is needed
For seeing them, with added elegance,
In red and white.

Such is the beauty of these mountains and rivers
That has been admired by unnumbered heroes—
The great emperors of Qin and Han
Lacking literary brilliance,
Those of Tang and Song
Having but few romantic inclinations,
And the prodigious Gengis Khan[1]
Knowing only how to bend his bow
    and shoot at vultures.
All are past and gone!
For men of vision
We must seek among the present generation.

---

*The highlands are those of Shaanxi and Shanxi.
[1]*Gengis Khan:* Genghis Khan (ca. 1162–1227), Mongol conqueror who became emperor of China

6

The Chinese People Have Stood Up
September 1949

By September 1949, the CCP had clearly defeated the GMD and set about establishing its new government, the People's Republic of China (PRC). Part of that system, drawing on the promise of a united front of Communists and non-Communists made in Mao's 1940 "On New Democracy" (see Document 2), was the formation of the Chinese Political Consultative Congress, which brought together a range of left-wing and independent parties willing to work with the new CCP government. This selection comes from Mao's speech at the preparatory session of the Consultative Congress. The tone of the speech is celebratory. After decades of turmoil and the near eradication of the CCP, victory was at hand. Mao's themes are nationalist rather than socialist; he dwells on China's new prestige and its national success in throwing off the Japanese, the Europeans (referred to as imperialists), and the corrupt government of the GMD. This was the high point of Mao's popular prestige, because he was widely seen as the savior who had delivered China's hard-earned independence and stature. Yet Mao hints at the continuing need to maintain "vigilance" against unnamed "reactionaries," leaving the door open for the terrible political purges and mass campaigns of the future.

This talk is the source of one of Mao's most remembered—and misremembered—statements: "We [the Chinese people] have stood up." Popular memory has it that Mao said this from the top of Tiananmen Gate on October 1, 1949, but he actually said it at this meeting in September. The title here comes from the editors of the most complete Chinese edition of Mao's writing, published in the 1990s and thus reflecting the continued power of the words.

"Zhongguo ren congci zhanqi laile," in Mao Zedong wenji (Beijing: Renmin chubanshe, 1996), 5:342–46, based on the People's Daily version of September 22, 1949. Translation from Stuart Schram, The Political Thought of Mao Tse-tung, 2nd ed. (New York: Praeger, 1969), 167–68.

... Our conference is one of great nationwide popular unity.

Such great nationwide popular unity has been achieved because we have vanquished the Guomindang reactionary government, which is aided by American imperialism. In the course of little more than three years, the heroic Chinese People's Liberation Army, an army such as the world has seldom seen, crushed the offensive of the several million troops of the American-supported Guomindang reactionary government, thereby enabling us to swing over to the counter-offensive and the offensive. . . .

We have a common feeling that our work will be recorded in the history of mankind, and that it will clearly demonstrate that the Chinese, who comprise one quarter of humanity, have begun to stand up. The Chinese have always been a great, courageous, and industrious people. It was only in modern times that they have fallen behind, and this was due solely to the oppression and exploitation of foreign imperialism and the domestic reactionary government.

For more than a century, our predecessors never paused in their indomitable struggles against the foreign and domestic oppressors. These struggles include the Revolution of 1911, led by Sun Yat-sen, the great pioneer of China's revolution. Our predecessors instructed us to carry their work to completion. We are doing this now. We have united ourselves and defeated both our foreign and domestic oppressors by means of the people's liberation war and the people's great revolution, and we proclaim the establishment of the People's Republic of China.

Henceforth, our nation will enter the large family of peace-loving and freedom-loving nations of the world. It will work bravely and industriously to create its own civilization and happiness, and will, at the same time, promote world peace and freedom. Our nation will never again be an insulted nation. We have stood up. Our revolution has gained the sympathy and acclamation of the broad masses throughout the world. We have friends the world over.

Our revolutionary work is not yet concluded. . . . The imperialists and the domestic reactionaries will certainly not take their defeat lying down. . . . Daily, hourly, they will try to restore their rule in China. . . . We must not relax our vigilance. . . .

The people's democratic dictatorship and unity with international friends will enable us to obtain rapid success in our construction work. . . . Our population of 475 million and our national territory of 9,597 million square kilometres are factors in our favour. It is true that there are difficulties ahead of us, a great many of them. But we firmly believe that all the difficulties will be surmounted by the heroic struggle of all the people of our country. The Chinese people has had ample experience in overcoming difficulties. If we and our predecessors could come through the long period of extreme difficulties and defeat the powerful domestic and foreign reactionaries, why can we not build up a prosperous and flourishing country after our victory? . . .

An upsurge in cultural construction will inevitably follow in the wake of the upsurge of economic construction. The era in which the Chinese were regarded as uncivilized is now over. We will emerge in the world as a nation with a high culture.

Our national defence will be consolidated and no imperialist will be allowed to invade our territory again. . . .

Let the domestic and foreign reactionaries tremble before us. Let them say that we are no good at this and no good at that. Through the Chinese people's indomitable endeavours, we will steadily reach our goal.

7

On the Correct Handling of Contradictions among the People

June 1957

This is arguably Mao Zedong's most important single contribution to the development of state socialism and Marxist-Leninist theory after 1949. Here Mao confronts directly the ultimately fatal flaw of the Leninist state: the lack of political feedback (such as that provided by secret-ballot elections in liberal democracies). Since the CCP claims access to truth as the scientific voice of the working class, to oppose the CCP is to oppose all that is good and right. Yet everyone, including the CCP leadership, knew that the CCP was not infallible despite its formal claims. In our terms, if a government does not find a way to get accurate feedback on how its policies work in practice, it will soon crash—like a pilot

"Guanyu zhengque chuli renmin neibu maodun de wenti," Renmin ribao, 19 June 1957. Translation from Selected Works of Mao Tse-tung (Peking: Foreign Languages Press, 1977), 5:384–421.

flying blind. Mao set out in February 1957 to solve this problem for China.
This is the most philosophical of Mao's writings included in this reader. His line of argument, especially in the first several pages, is a model of his application of Marxist dialectics to a pressing political problem. Thus he concludes his first review of the economy by saying, "The ceaseless emergence and ceaseless resolution of contradictions constitute the dialectical law of the development of things." Throughout the text, Mao analyzes the whole range of social issues in China—from agriculture to industry to intellectuals to foreign policy—in these philosophical terms. He begins his practical consideration of "contradictions among the people" by dispensing with counterrevolutionaries (subjects, of course, of antagonistic contradictions). He then turns to major social groups in China: various agrarian classes (poor, middle, and rich peasants), businessmen and industrialists, and intellectuals. Mao and the CCP needed the talents of all these social classes, but especially the technical skills of China's intellectual elite, to move China's economic modernization program forward.
Mao offered this comprehensive philosophical and political blueprint at a time of challenge for the new Communist government of China. The first seven years of CCP had gone quite well, and Mao was the revered leader of the nation. But now there were problems. De-Stalinization in the USSR and anti-Soviet rebellions in Communist Poland and Hungary in 1956 added urgency to Mao's conviction that the CCP had succumbed to bureaucratism—rule by a heartless and all-powerful state administration.
Mao's solution was to give the party a dose of self and mutual criticism, one of the techniques of rectification study from the Yan'an period, but with a twist—he would ask a new part of the "masses," China's literate public and intellectuals, to improve the party through public criticism in the press. This required an explanation, in terms of Marxist theory, of why it might be permissible to criticize "the voice of the people." Mao provides that theoretical development in this speech. He posits two kinds of contradictions—antagonistic ones between "us" and the enemy (such as relations between China and the United States) and non-antagonistic ones within the ranks of the people.
It was a brilliant idea, and one that promised to make the Communist government more open to public opinion and more aware of the trade-offs necessary in administering a modernizing nation. Indeed, in the February 1957 version of the speech, Mao essentially guaranteed

immunity from political prosecution if professionals and intellectuals (in several nominal and powerless "democratic parties") would speak up.[1]

The chronology of the spring of 1957 is covered in the introduction. When the public "blooming" Mao called for finally occurred in May, it was much more harshly critical of the CCP than he had anticipated. The party promptly shut down these criticisms, and the Anti-Rightist Campaign started in late June to prosecute those who had spoken up. The betrayal of Mao's promise to intellectuals was embodied in the published version of "On the Correct Handling of Contradictions" on June 19 in the People's Daily, *the CCP's national newspaper. There it appeared in a highly edited form minus the guarantees of immunity and with a new set of six criteria "for distinguishing fragrant flowers from poisonous weeds" that made the May critics guilty after the fact. Thus the politics around the text negated its promising content. Mao's practice was out of step with his pronouncements.*

This translation is the official Selected Works *version based on the* People's Daily *article in June 1957. This is the version approved by the party as a "collective wisdom" edition (see "A Note about the Texts"), and it is the version studied by millions of Chinese, even though it is not the version Mao first presented in February 1957. The revised version is important for more than the deletion of Mao's promise to respect all public criticism. The "speaking notes" version from February is rambling and, at points, incoherent. Party editors and Mao clearly put a great deal of effort into systematizing Mao's thoughts on the nature of contradictions in this public version.*

Our general subject is the correct handling of contradictions among the people. For convenience, let us discuss it under twelve subheadings. Although reference will be made to contradictions between ourselves and the enemy, this discussion will center on contradictions among the people.

[1]The "speaking notes" of Mao's February 27, 1957, "Contradictions" talk is translated in Roderick MacFarquhar, Timothy Cheek, and Eugene Wu, eds., *The Secret Speeches of Chairman Mao: From the Hundred Flowers to the Great Leap Forward* (Cambridge: Harvard Contemporary China Series, 1989), 131–90.

I. TWO TYPES OF CONTRADICTIONS DIFFERING IN NATURE

Never before has our country been as united as it is today. The victories of the bourgeois-democratic revolution and of the socialist revolution and our achievements in socialist construction have rapidly changed the face of the old China. A still brighter future lies ahead for our motherland. The days of national disunity and chaos which the people detested are gone, never to return. Led by the working class and the Communist Party, our 600 million people, united as one, are engaged in the great task of building socialism. The unification of our country, the unity of our people and the unity of our various nationalities—these are the basic guarantees for the sure triumph of our cause. However, this does not mean that contradictions no longer exist in our society. To imagine that none exist is a naive idea which is at variance with objective reality. We are confronted with two types of social contradictions—those between ourselves and the enemy and those among the people. The two are totally different in nature.

To understand these two different types of contradictions correctly, we must first be clear on what is meant by "the people" and what is meant by "the enemy."... At the present stage, the period of building socialism, the classes, strata and social groups which favour, support and work for the cause of socialist construction all come within the category of the people, while the social forces and groups which resist the socialist revolution and are hostile to or sabotage socialist construction are all enemies of the people.

The contradictions between ourselves and the enemy are antagonistic contradictions. Within the ranks of the people, the contradictions among the working people are non-antagonistic, while those between the exploited and the exploiting classes have a non-antagonistic as well as an antagonistic aspect. There have always been contradictions among the people, but they are different in content in each period of the revolution and in the period of building socialism. In the conditions prevailing in China today, the contradictions among the people comprise the contradictions within the working class, the contradictions within the peasantry, the contradictions within the intelligentsia, the contradictions between the working class and the peasantry, the contradictions between the workers and peasants on the one hand and the intellectuals on the other, the contradictions between the working class and other sections of the working people on the one hand and the national bourgeoisie on the other, the contradictions within the national bourgeoisie, and so on. Our People's Government is one that genuinely represents the people's interests, it is a government that serves the people. Nevertheless, there are still certain contradictions between this government and the people. These include the contradictions between the interests of the state and the interests of the collective on the one hand and the interests of the individual on the other, between democracy and centralism, between the leadership and the led, and the contradictions arising from the bureaucratic style of work of some of the state personnel in their relations with the masses. All these are also contradictions among the people. Generally speaking, the fundamental identity of the people's interests underlies the contradictions among the people.

In our country, the contradiction between the working class and the national bourgeoisie comes under the category of contradictions among the people. By and large, the class struggle between the two is a class struggle within the ranks of the people, because the Chinese national bourgeoisie has a dual character. In the period of the bourgeois-democratic revolution, it had both a revolutionary and a conciliationist side to its character. In the period of the socialist revolution, exploitation of the working class for profit constitutes one side of the character of the national bourgeoisie, while its support of the Constitution and its willingness to accept socialist transformation constitute the other. The national bourgeoisie differs from the imperialists, the landlords and the bureaucrat-capitalists. The contradiction between the national bourgeoisie and the working class is one between exploiter and exploited, and is by nature antagonistic. But in the concrete conditions of China, this antagonistic contradiction between the two classes, if properly handled, can be transformed into a non-antagonistic one and be resolved by peaceful methods. However, the contradiction between the working class and the national bourgeoisie will change into a contradiction between ourselves and the enemy if we do not handle it properly and do not follow the policy of uniting with, criticizing and educating the national bourgeoisie, or if the national bourgeoisie does not accept this policy of ours.

Since they are different in nature, the contradictions between ourselves and the enemy and the contradictions among the people must be resolved by different methods. To put it briefly, the former entail drawing a clear distinction between ourselves and the enemy, and the latter entail drawing a clear distinction between right and wrong. It is of course true that the distinction between ourselves and the enemy is also one of right and wrong. For example, the question of who is in the right, we or the domestic and foreign reactionaries, the imperialists, the feudalists and bureaucrat-capitalists, is also one of right and wrong, but it is in a different category from questions of right and wrong among the people.

Our state is a people's democratic dictatorship led by the working class and based on the worker-peasant alliance. What is this dictatorship for? Its first function is internal, namely, to suppress the reactionary classes and elements and those exploiters who resist the socialist revolution, to suppress those who try to wreck our socialist construction, or in other words, to resolve the contradictions between ourselves and the internal enemy. For instance, to arrest, try and sentence certain counter-revolutionaries, and to deprive landlords and bureaucrat-capitalists of their right to vote and their freedom of speech for a certain period of time—all this comes within the scope of our dictatorship. To maintain public order and safeguard the interests of the people, it is necessary to exercise dictatorship as well over thieves, swindlers, murderers, arsonists, criminal gangs and other scoundrels who seriously disrupt public order. The second function of this dictatorship is to protect our country from subversion and possible aggression by external enemies. In such contingencies, it is the task of this dictatorship to resolve the contradiction between ourselves and the external enemy. The aim of this dictatorship is to protect all our people so that they can devote themselves to peaceful labour and make China a socialist country with modern industry, modern agriculture, and modern science and culture. Who is to exercise this dictatorship? Naturally, the working class and the entire people under its leadership. Dictatorship does not apply within the ranks of the people. The people cannot exercise dictatorship over themselves, nor must one section of the people oppress another. Law-breakers among the people will be punished according to law, but this is different in principle from the exercise of dictatorship to suppress enemies of the people. What applies among the people is democratic centralism. Our Constitution lays it down that citizens of the People's Republic of China enjoy freedom of speech, the press, assembly, association, procession, demonstration, religious belief, and so on. Our Constitution also provides that the organs of state must practise democratic centralism, that they must rely on the masses and that their personnel must serve the people. Our socialist democracy is the broadest kind of democracy, such as is not to be found in any bourgeois state. Our dictatorship is the people's democratic dictatorship led by the working class and based on the worker-peasant alliance. That is to say, democracy operates within the ranks of the people, while the working class, uniting with all others enjoying civil rights, and in the first place with the peasantry, enforces dictatorship over the reactionary classes and elements and all those who resist socialist transformation and oppose socialist construction. By civil rights, we mean, politically, the rights of freedom and democracy.

But this freedom is freedom with leadership and this democracy is democracy under centralized guidance, not anarchy. Anarchy does not accord with the interests or wishes of the people.

Certain people in our country were delighted by the Hungarian incident.[2] They hoped that something similar would happen in China, that thousands upon thousands of people would take to the streets to demonstrate against the People's Government. Their hopes ran counter to the interests of the masses and therefore could not possibly win their support. Deceived by domestic and foreign counter-revolutionaries, a section of the people in Hungary made the mistake of resorting to violence against the people's government, with the result that both the state and the people suffered. The damage done to the country's economy in a few weeks of rioting will take a long time to repair. In our country there were some others who wavered on the question of the Hungarian incident because they were ignorant of the real state of affairs in the world. They think that there is too little free-dom under our people's democracy and that there is more freedom under Western parliamentary democracy. They ask for a two-party system as in the West, with one party in office and the other in oppo-sition. But this so-called two-party system is nothing but a device for maintaining the dictatorship of the bourgeoisie; it can never guarantee freedoms to the working people. As a matter of fact, freedom and democracy exist not in the abstract, but only in the concrete. In a soci-ety where class struggle exists, if there is freedom for the exploiting classes to exploit the working people, there is no freedom for the working people not to be exploited. If there is democracy for the bour-geoisie, there is no democracy for the proletariat and other working people. The legal existence of the Communist Party is tolerated in some capitalist countries, but only to the extent that it does not endan-ger the fundamental interests of the bourgeoisie; it is not tolerated beyond that. Those who demand freedom and democracy in the

[2]*Hungarian incident:* In October 1956, reformist leaders in socialist Hungary, mobi-lized by student protests, expelled Stalinist leaders and declared military independence from the Soviet Union. The Soviets quickly crushed the reform government and reasserted control over Hungary by military force.

abstract regard democracy as an end and not as a means. Democracy as such sometimes seems to be an end, but it is in fact only a means. Marxism teaches us that democracy is part of the superstructure and belongs to the realm of politics. That is to say, in the last analysis, it serves the economic base. The same is true of freedom. Both democracy and freedom are relative, not absolute, and they come into being and develop in specific historical conditions. Within the ranks of the people, democracy is correlative with centralism and freedom with discipline. They are the two opposites of a single entity, contradictory as well as united, and we should not one-sidedly emphasize one to the exclusion of the other. Within the ranks of the people, we cannot do without freedom, nor can we do without discipline; we cannot do without democracy, nor can we do without centralism. This unity of democracy and centralism, of freedom and discipline, constitutes our democratic centralism. Under this system, the people enjoy broad democracy and freedom, but at the same time they have to keep within the bounds of socialist discipline. All this is well understood by the masses.

In advocating freedom with leadership and democracy under centralized guidance, we in no way mean that coercive measures should be taken to settle ideological questions or questions involving the distinction between right and wrong among the people. All attempts to use administrative orders or coercive measures to settle ideological questions or questions of right and wrong are not only ineffective but harmful. We cannot abolish religion by administrative order or force people not to believe in it. We cannot compel people to give up idealism, any more than we can force them to embrace Marxism. The only way to settle questions of an ideological nature or controversial issues among the people is by the democratic method, the method of discussion, criticism, persuasion and education, and not by the method of coercion or repression. To be able to carry on their production and studies effectively and to lead their lives in peace and order, the people want their government and those in charge of production and of cultural and educational organizations to issue appropriate administrative regulations of an obligatory nature. It is common sense that without them the maintenance of public order would be impossible. Administrative regulations and the method of persuasion and education complement each other in resolving contradictions among the people. In fact, administrative regulations for the maintenance of public order must be accompanied by persuasion and education, for in many cases regulations alone will not work.

This democratic method of resolving contradictions among the people was epitomized in 1942 in the formula "unity—criticism—unity." To elaborate, that means starting from the desire for unity, resolving contradictions through criticism or struggle, and arriving at a new unity on a new basis. In our experience this is the correct method of resolving contradictions among the people. In 1942 we used it to resolve contradictions inside the Communist Party, namely, the contradictions between the dogmatists and the great majority of the membership, and between dogmatism and Marxism. The "Left" dogmatists had resorted to the method of "ruthless struggle and merciless blows" in inner-Party struggle. It was the wrong method. In criticizing "Left" dogmatism, we did not use this old method but adopted a new one, that is, one of starting from the desire for unity, distinguishing between right and wrong through criticism or struggle, and arriving at a new unity on a new basis. This was the method used in the rectification movement of 1942. Within a few years, by the time the Chinese Communist Party held its Seventh National Congress in 1945, unity was achieved throughout the Party as anticipated, and consequently the people's revolution triumphed. Here, the essential thing is to start from the desire for unity. For without this desire for unity, the struggle, once begun, is certain to throw things into confusion and get out of hand. Wouldn't this be the same as "ruthless struggle and merciless blows"? And what Party unity would there be left? It was precisely this experience that led us to the formula "unity—criticism—unity." Or, in other words, "learn from past mistakes to avoid future ones and cure the sickness to save the patient." We extended this method beyond our Party. We applied it with great success in the anti-Japanese base areas in dealing with the relations between the leadership and the masses, between the army and the people, between officers and men, between the different units of the army, and between the different groups of cadres. The use of this method can be traced back to still earlier times in our Party's history. Ever since 1927 when we built our revolutionary armed forces and base areas in the south, this method had been used to deal with the relations between the Party and the masses, between the army and the people, between officers and men, and with other relations among the people. The only difference was that during the anti-Japanese war we employed this method much more consciously. And since the liberation of the whole country, we have employed this same method of "unity—criticism—unity" in our relations with the democratic parties[3] and with industrial and commercial circles. Our task now is to continue to extend and make still better use of this method throughout the ranks of the people; we want all our factories, co-operatives, shops, schools, offices and people's organizations, in a word, all our 600 million people, to use it in resolving contradictions among themselves.

In ordinary circumstances, contradictions among the people are not antagonistic. But if they are not handled properly, or if we relax our vigilance and lower our guard, antagonism may arise. In a socialist country, a development of this kind is usually only a localized and temporary phenomenon. The reason is that the system of exploitation of man by man has been abolished and the interests of the people are fundamentally identical. The antagonistic actions which took place on a fairly wide scale during the Hungarian incident were the result of the operations of both domestic and foreign counter-revolutionary elements. This was a particular as well as a temporary phenomenon. It was a case of the reactionaries inside a socialist country, in league with the imperialists, attempting to achieve their conspiratorial aims by taking advantage of contradictions among the people to foment dissension and stir up disorder. The lesson of the Hungarian incident merits attention.

Many people seem to think that the use of the democratic method to resolve contradictions among the people is something new. Actually it is not. Marxists have always held that the cause of the proletariat must depend on the masses of the people and that Communists must use the democratic method of persuasion and education when working among the labouring people and must on no account resort to commandism or coercion. The Chinese Communist Party faithfully adheres to this Marxist-Leninist principle. It has been our consistent view that under the people's democratic dictatorship two different methods, one dictatorial and the other democratic, should be used to resolve the two types of contradictions which differ in nature—those between ourselves and the enemy and those among the people. This idea has been explained again and again in many Party documents and in speeches by many leading comrades of our Party. In my article "On the People's Democratic Dictatorship," written in 1949, I said, "The combination of these two aspects, democracy for the people and

---

[3] *democratic parties:* in the 1950s, remnants of truly independent political parties active before 1949. Under the Communist government, they served more as interest groups for professionals. They were retained by the CCP as reminders of the promise of New Democracy, and here Mao is trying to use that legitimacy to reform the CCP.

dictatorship over the reactionaries, is the people's democratic dictatorship." I also pointed out that in order to settle problems within the ranks of the people "the method we employ is democratic, the method of persuasion, not of compulsion." Again, in addressing the Second Session of the First National Committee of the Political Consultative Conference in June 1950, I said:

> The people's democratic dictatorship uses two methods. Towards the enemy, it uses the method of dictatorship, that is, for as long a period of time as is necessary it does not permit them to take part in political activity and compels them to obey the law of the People's Government, to engage in labour and, through such labour, be transformed into new men. Towards the people, on the contrary, it uses the method of democracy and not of compulsion, that is, it must necessarily let them take part in political activity and does not compel them to do this or that but uses the method of democracy to educate and persuade. Such education is self-education for the people, and its basic method is criticism and self-criticism.

Thus, on many occasions we have discussed the use of the democratic method for resolving contradictions among the people; furthermore, we have in the main applied it in our work, and many cadres and many other people are familiar with it in practice. Why then do some people now feel that it is a new issue? Because, in the past, the struggle between ourselves and the enemy, both internal and external, was most acute, and contradictions among the people therefore did not attract as much attention as they do today.

Quite a few people fail to make a clear distinction between these two different types of contradictions—those between ourselves and the enemy and those among the people—and are prone to confuse the two. It must be admitted that it is sometimes quite easy to do so. We have had instances of such confusion in our work in the past. In the course of cleaning out counter-revolutionaries good people were sometimes mistaken for bad, and such things still happen today. We are able to keep mistakes within bounds because it has been our policy to draw a sharp line between ourselves and the enemy and to rectify mistakes whenever discovered.

Marxist philosophy holds that the law of the unity of opposites is the fundamental law of the universe. This law operates universally, whether in the natural world, in human society, or in man's thinking. Between the opposites in a contradiction there is at once unity and struggle, and it is this that impels things to move and change.

Contradictions exist everywhere, but their nature differs in accordance with the different nature of different things. In any given thing, the unity of opposites is conditional, temporary and transitory, and hence relative, whereas the struggle of opposites is absolute. Lenin gave a very clear exposition of this law. It has come to be understood by a growing number of people in our country. But for many people it is one thing to accept this law and quite another to apply it in examining and dealing with problems. Many dare not openly admit that contradictions still exist among the people of our country, while it is precisely these contradictions that are pushing our society forward. Many do not admit that contradictions still exist in socialist society, with the result that they become irresolute and passive when confronted with social contradictions; they do not understand that socialist society grows more united and consolidated through the ceaseless process of correctly handling and resolving contradictions. For this reason, we need to explain things to our people, and to our cadres in the first place, in order to help them understand the contradictions in socialist society and learn to use correct methods for handling them.

Contradictions in socialist society are fundamentally different from those in the old societies, such as capitalist society. In capitalist society contradictions find expression in acute antagonisms and conflicts, in sharp class struggle; they cannot be resolved by the capitalist system itself and can only be resolved by socialist revolution. The case is quite different with contradictions in socialist society; on the contrary, they are not antagonistic and can be ceaselessly resolved by the socialist system itself.

In socialist society the basic contradictions are still those between the relations of production and the productive forces and between the superstructure and the economic base.[4] However, they are fundamentally different in character and have different features from the contradictions between the relations of production and the productive forces and between the superstructure and the economic base in the old societies. The present social system of our country is far superior to that of the old days. If it were not so, the old system would not have been overthrown and the new system could not have been established. In saying that the socialist relations of production correspond

[4]Mao uses the terms *relations of production* and *productive forces* to express the Marxist view of who owns the factories, farms, and businesses and how that form of ownership relates to how things are made—that is, the level of modernization of industry and agriculture. Similarly, he uses *superstructure* and *economic base* to talk about the relationship between the economy and politics.

better to the character of the productive forces than did the old relations of production, we mean that they allow the productive forces to develop at a speed unattainable in the old society, so that production can expand steadily and increasingly meet the constantly growing needs of the people. Under the rule of imperialism, feudalism and bureaucrat-capitalism, the productive forces of the old China grew very slowly. For more than fifty years before liberation, China produced only a few tens of thousands of tons of steel a year, not counting the output of the northeastern provinces. If these provinces are included, the peak annual steel output only amounted to a little over 900,000 tons. In 1949, the national steel output was a little over 100,000 tons. Yet now, a mere seven years after the liberation of our country, steel output already exceeds 4,000,000 tons. In the old China, there was hardly any machine-building industry, to say nothing of the automobile and aircraft industries; now we have all three. When the people overthrew the rule of imperialism, feudalism and bureaucrat-capitalism, many were not clear as to which way China should head— towards capitalism or towards socialism. Facts have now provided the answer: Only socialism can save China. The socialist system has promoted the rapid development of the productive forces of our country, a fact even our enemies abroad have had to acknowledge.

But our socialist system has only just been set up; it is not yet fully established or fully consolidated. In joint state-private industrial and commercial enterprises, capitalists still get a fixed rate of interest on their capital, that is to say, exploitation still exists. So far as ownership is concerned, these enterprises are not yet completely socialist in nature. A number of our agricultural and handicraft producers' co-operatives are still semi-socialist, while even in the fully socialist co-operatives certain specific problems of ownership remain to be solved. Relations between production and exchange in accordance with socialist principles are being gradually established within and between all branches of our economy, and more and more appropriate forms are being sought. . . . To sum up, socialist relations of production have been established and are in correspondence with the growth of the productive forces, but these relations are still far from perfect, and this imperfection stands in contradiction to the growth of the productive forces. Apart from correspondence as well as contradiction between the relations of production and the growth of the productive forces, there is correspondence as well as contradiction between the superstructure and the economic base. The superstructure, comprising the state system and laws of the people's democratic dictatorship and the socialist ideology guided by Marxism-Leninism, plays a positive role in facilitating the victory of socialist transformation and the socialist way of organizing labour; it is in correspondence with the socialist economic base, that is, with socialist relations of production. But the existence of bourgeois ideology, a certain bureaucratic style of work in our state organs and defects in some of the links in our state institutions are in contradiction with the socialist economic base. We must continue to resolve all such contradictions in the light of our specific conditions. Of course, new problems will emerge as these contradictions are resolved. And further efforts will be required to resolve the new contradictions. For instance, a constant process of readjustment through state planning is needed to deal with the contradiction between production and the needs of society, which will long remain an objective reality. Every year our country draws up an economic plan in order to establish a proper ratio between accumulation and consumption and achieve an equilibrium between production and needs. Equilibrium is nothing but a temporary, relative, unity of opposites. By the end of each year, this equilibrium, taken as a whole, is upset by the struggle of opposites; the unity undergoes a change, equilibrium becomes disequilibrium, unity becomes disunity, and once again it is necessary to work out an equilibrium and unity for the next year. Herein lies the superiority of our planned economy. As a matter of fact, this equilibrium, this unity, is partially upset every month or every quarter, and partial readjustments are called for. Sometimes, contradictions arise and the equilibrium is upset because our subjective arrangements do not conform to objective reality; this is what we call making a mistake. The ceaseless emergence and ceaseless resolution of contradictions constitute the dialectical law of the development of things.

Today, matters stand as follows. The large-scale, turbulent class struggles of the masses characteristic of times of revolution have in the main come to an end, but class struggle is by no means entirely over. While welcoming the new system, the masses are not yet quite accustomed to it. Government personnel are not sufficiently experienced and have to undertake further study and investigation of specific policies. In other words, time is needed for our socialist system to become established and consolidated, for the masses to become accustomed to the new system, and for government personnel to learn and acquire experience. It is therefore imperative for us at this juncture to raise the question of distinguishing contradictions among the people from those between ourselves and the enemy, as well as the question of the correct handling of contradictions among the people, in order to unite the people of all nationalities in our country for the new battle, the battle against nature, develop our economy and culture, help the whole nation to traverse this period of transition relatively smoothly, consolidate our new system and build up our new state.

II. THE QUESTION OF ELIMINATING COUNTER-REVOLUTIONARIES

The elimination of counter-revolutionaries is a struggle of opposites as between ourselves and the enemy. Among the people, there are some who see this question in a somewhat different light. Two kinds of people hold views differing from ours. Those with a Right deviation in their thinking make no distinction between ourselves and the enemy and take the enemy for our own people. They regard as friends the very persons whom the masses regard as enemies. Those with a "Left" deviation in their thinking magnify contradictions between ourselves and the enemy to such an extent that they take certain contradictions among the people for contradictions with the enemy and regard as counter-revolutionaries persons who are actually not. Both these views are wrong. Neither makes possible the correct handling of the problem of eliminating counter-revolutionaries or a correct assessment of this work.

To form a correct evaluation of our work in eliminating counter-revolutionaries, let us see what repercussions the Hungarian incident has had in China. After its occurrence there was some unrest among a section of our intellectuals, but there were no squalls. Why? One reason, it must be said, was our success in eliminating counter-revolutionaries fairly thoroughly.

Of course, the consolidation of our state is not due primarily to the elimination of counter-revolutionaries. It is due primarily to the fact that we have a Communist Party and a Liberation Army both tempered in decades of revolutionary struggle, and a working people likewise so tempered. Our Party and our armed forces are rooted in the masses, have been tempered in the flames of a protracted revolution and have the capacity to fight. Our People's Republic was not built overnight, but developed step by step out of the revolutionary base areas. A number of democratic personages have also been tempered in the struggle in varying degrees, and they have gone through troubled times together with us. Some intellectuals were tempered in the struggles against imperialism and reaction; since liberation many have gone through a process of ideological remoulding[5] aimed at enabling them to distinguish clearly between ourselves and the enemy. In addition, the consolidation of our state is due to the fact that our economic measures are basically sound, that the people's life is secure and steadily improving, that our policies towards the national bourgeoisie and other classes are correct, and so on. Nevertheless, our success in eliminating counter-revolutionaries is undoubtedly an important reason for the consolidation of our state. For all these reasons, with few exceptions our college students are patriotic and support socialism and did not give way to unrest during the Hungarian incident, even though many of them come from families of non-working people. The same was true of the national bourgeoisie, to say nothing of the basic masses—the workers and peasants.

After liberation, we rooted out a number of counter-revolutionaries. Some were sentenced to death for major crimes. This was absolutely necessary, it was the demand of the masses, and it was done to free them from long years of oppression by the counter-revolutionaries and all kinds of local tyrants, in other words, to liberate the productive forces. If we had not done so, the masses would not have been able to lift their heads. . . .

In our work of eliminating counter-revolutionaries successes were the main thing, but there were also mistakes. In some cases there were excesses and in others counter-revolutionaries slipped through our net. Our policy is: "Counter-revolutionaries must be eliminated wherever found, mistakes must be corrected whenever discovered." Our line in the work of eliminating counter-revolutionaries is the mass line. Of course, even with the mass line mistakes may still occur, but they will be fewer and easier to correct. The masses gain experience through struggle. From the things done correctly they gain the experience of how things are done correctly. From the mistakes made they gain the experience of how mistakes are made. . . .

The present situation with regard to counter-revolutionaries can be described in these words: There still are counter-revolutionaries, but not many. In the first place, there still are counter-revolutionaries.

---

[5]*ideological remoulding:* one of the core aspects of Maoism, this is the process of changing the way a person thinks, known in the West as brainwashing. Drawing on the ideological study sessions of Yan'an, this effort by the CCP was an orchestrated way to effect the change in class attitude Mao talked about in "Talks at the Yan'an Conference on Literature and Art" (see Document 3).

Some people say that there aren't any more left and all is well and that we can therefore lay our heads on our pillows and just drop off to sleep. But this is not the way things are. The fact is, there still are counter-revolutionaries (of course, that is not to say you'll find them everywhere and in every organization), and we must continue to fight them. It must be understood that the hidden counter-revolutionaries still at large will not take things lying down, but will certainly seize every opportunity to make trouble. The U.S. imperialists and the Chiang Kai-shek clique are constantly sending in secret agents to carry on disruptive activities. Even after all the existing counter-revolutionaries have been combed out, new ones are likely to emerge. If we drop our guard, we shall be badly fooled and shall suffer severely. Counter-revolutionaries must be rooted out with a firm hand wherever they are found making trouble. . . .

III. THE QUESTION OF THE CO-OPERATIVE TRANSFORMATION OF AGRICULTURE

We have a rural population of over 500 million, so how our peasants fare has a most important bearing on the development of our economy and the consolidation of our state power. In my view, the situation is basically sound. The co-operative transformation of agriculture has been successfully accomplished,[6] and this has resolved the great contradiction in our country between socialist industrialization and the individual peasant economy. As the co-operative transformation of agriculture was completed so rapidly, some people were worried and wondered whether something untoward might occur. There are indeed some faults, but fortunately they are not serious and on the whole the movement is healthy. The peasants are working with a will, and last year there was an increase in the country's grain output despite the worst floods, droughts and gales in years. Now there are people who are stirring up a miniature typhoon, they are saying that co-operation is no good, that there is nothing superior about it. Is co-operation superior or not? Among the documents distributed at today's meeting there is one about the Wang Guofan Co-operative in Zunhua County,

[6]*Agricultural cooperativization* was the CCP-sponsored movement to bring the newly independent farmers of China into closer economic cooperation. This process began in the mid-1950s but took off, with Mao's encouragement, in 1956. By 1957 Mao was thinking of the next step, which would become the people's communes of the Great Leap Forward in 1958.

Hebei Province, which I suggest you read. This co-operative is situated in a hilly region which was very poor in the past and which for a number of years depended on relief grain from the People's Government. When the co-operative was first set up in 1953, people called it the "paupers' co-op." But it has become better off year by year, and now, after four years of hard struggle, most of its households have reserves of grain. What was possible for this co-operative should also be possible for others to achieve under normal conditions in the same length of time or a little longer. Clearly there are no grounds for saying that something has gone wrong with agricultural co-operation.

It is also clear that it takes hard struggle to build co-operatives. New things always have to experience difficulties and setbacks as they grow. It is sheer fantasy to imagine that the cause of socialism is all plain sailing and easy success, with no difficulties and setbacks, or without the exertion of tremendous efforts.

Who are the active supporters of the co-operatives? The overwhelming majority of the poor and lower-middle peasants who constitute more than 70 per cent of the rural population. Most of the other peasants are also placing their hopes on the co-operatives. Only a very small minority are really dissatisfied. . . .

The co-operatives are now in the process of gradual consolidation. There are certain contradictions that remain to be resolved, such as those between the state and the co-operatives and those in and between the co-operatives themselves.

To resolve these contradictions we must pay constant attention to the problems of production and distribution. On the question of production, the co-operative economy must be subject to the unified economic planning of the state, while retaining a certain flexibility and independence that do not run counter to the state's unified plan or its policies, laws and regulations. At the same time, every household in a co-operative must comply with the over-all plan of the co-operative or production team to which it belongs, though it may make its own appropriate plans in regard to land allotted for personal needs and to other individually operated economic undertakings. On the question of distribution, we must take the interests of the state, the collective and the individual into account. We must properly handle the three-way relationship between the state agricultural tax, the co-operative's accumulation fund and the peasants' personal income, and take constant care to make readjustments so as to resolve contradictions between them. Accumulation is essential for both the state and the co-operative, but in neither case should it be excessive. We should do everything possible to enable the peasants in normal years to raise their personal incomes annually through increased production....

IV. THE QUESTION OF THE INDUSTRIALISTS AND BUSINESSMEN

With regard to the transformation of our social system, the year 1956 saw the conversion of privately owned industrial and commercial enterprises into joint state-private enterprises as well as the co-operative transformation of agriculture and handicrafts. The speed and smoothness of this conversion were closely bound up with our treating the contradiction between the working class and the national bourgeoisie as a contradiction among the people. Has this class contradiction been completely resolved? No, not yet. That will take a considerable period of time. However, some people say the capitalists have been so remoulded that they are now not very different from the workers and that further remoulding is unnecessary. Others go so far as to say that the capitalists are even better than the workers. Still others ask, if remoulding is necessary, why isn't it necessary for the working class? Are these opinions correct? Of course not.

In the building of a socialist society, everybody needs remoulding—the exploiters and also the working people. Who says it isn't necessary for the working class? Of course, the remoulding of the exploiters is essentially different from that of the working people, and the two must not be confused. The working class remoulds the whole of society in class struggle and in the struggle against nature, and in the process it remoulds itself. It must ceaselessly learn in the course of work, gradually overcome its shortcomings and never stop doing so. Take for example those of us present here. Many of us make some progress each year, that is to say, we are remoulding ourselves each year. For myself, I used to have all sorts of non-Marxist ideas, and it was only later that I embraced Marxism. I learned a little Marxism from books and took the first steps in remoulding my ideology, but it was mainly through taking part in class struggle over the years that I came to be remoulded. And if I am to make further progress, I must continue to learn, otherwise I shall lag behind. Can the capitalists be so good that they need no more remoulding?

Some people contend that the Chinese bourgeoisie no longer has two sides to its character, but only one side. Is this true? No. While members of the bourgeoisie have become administrative personnel in joint state-private enterprises and are being transformed from exploiters into working people living by their own labour, they still get a fixed rate of interest on their capital in the joint enterprises, that is, they have not yet cut themselves loose from the roots of exploitation. Between them and the working class there is still a considerable gap in ideology, sentiments and habits of life. How can it be said that they no longer have two sides to their character? Even when they stop receiving their fixed interest payments and the "bourgeois" label is removed, they will still need ideological remoulding for quite some time. If, as is alleged, the bourgeoisie no longer has a dual character, then the capitalists will no longer have the task of studying and of remoulding themselves.

It must be said that this view does not tally either with the actual situation of our industrialists and businessmen or with what most of them want. During the past few years, most of them have been willing to study and have made marked progress. As their thorough remoulding can be achieved only in the course of work, they should engage in labour together with the staff and workers in the enterprises and regard these enterprises as the chief places in which to remould themselves. But it is also important for them to change some of their old views through study. Such study should be on a voluntary basis. When they return to the enterprises after being in study groups for some weeks, many industrialists and businessmen find that they have more of a common language with the workers and the representatives of state ownership, and so there are better possibilities for working together. They know from personal experience that it is good for them to keep on studying and remoulding[7] themselves. The idea mentioned above that study and remoulding are not necessary reflects the views not of the majority of industrialists and businessmen but of only a small number.

V. THE QUESTION OF THE INTELLECTUALS

The contradictions within the ranks of the people in our country also find expression among the intellectuals. The several million intellectuals who worked for the old society have come to serve the new society, and the question that now arises is how they can fit in with the needs of the new society and how we can help them to do so. This, too, is a contradiction among the people.

Most of our intellectuals have made marked progress during the last seven years. They have shown they are in favour of the socialist

[7]*studying and remoulding:* ideological remolding in CCP-organized study groups

system. Many are diligently studying Marxism, and some have become communists. The latter, though at present small in number, are steadily increasing. Of course, there are still some intellectuals who are sceptical about socialism or do not approve of it, but they are a minority.

China needs the services of as many intellectuals as possible for the colossal task of building socialism. We should trust those who are really willing to serve the cause of socialism and should radically improve our relations with them and help them solve the problems requiring solution, so that they can give full play to their talents. Many of our comrades are not good at uniting with intellectuals. They are stiff in their attitude towards them, lack respect for their work and interfere in certain scientific and cultural matters where interference is unwarranted. We must do away with all such shortcomings.

Although large numbers of intellectuals have made progress, they should not be complacent. They must continue to remould themselves, gradually shed their bourgeois world outlook and acquire the proletarian, communist world outlook so that they can fully fit in with the needs of the new society and unite with the workers and peasants. The change in world outlook is fundamental, and up to now most of our intellectuals cannot be said to have accomplished it. We hope that they will continue to make progress and that in the course of work and study they will gradually acquire the communist world outlook, grasp Marxism-Leninism and become integrated with the workers and peasants. ... But a thorough change in world outlook takes a very long time, and we should spare no pains in helping them and must not be impatient. Actually, there are bound to be some who ideologically will always be reluctant to accept Marxism-Leninism and communism. We should not be too exacting in what we demand of them; as long as they comply with the requirements laid down by the state and engage in legitimate pursuits, we should let them have opportunities for suitable work.

Among students and intellectuals there has recently been a falling off in ideological and political work, and some unhealthy tendencies have appeared. Some people seem to think that there is no longer any need to concern themselves with politics or with the future of the motherland and the ideals of mankind. It seems as if Marxism, once all the rage, is currently not so much in fashion. To counter these tendencies, we must strengthen our ideological and political work. Both students and intellectuals should study hard. In addition to the study of their specialized subjects, they must make progress ideologically and politically, which means they should study Marxism, current events and politics. Not to have a correct political orientation is like not having a soul. The ideological remoulding in the past was necessary and has yielded positive results. But it was carried on in a somewhat rough-and-ready fashion and the feelings of some people were hurt—this was not good. We must avoid such shortcomings in future. All departments and organizations should shoulder their responsibilities for ideological and political work. This applies to the Communist Party, the Youth League,[8] government departments in charge of this work, and especially to heads of educational institutions and teachers. Our educational policy must enable everyone who receives an education to develop morally, intellectually and physically and become a worker with both socialist consciousness and culture. We must spread the idea of building our country through diligence and thrift. We must help all our young people to understand that ours is still a very poor country, that we cannot change this situation radically in a short time, and that only through decades of united effort by our younger generation and all our people, working with their own hands, can China be made prosperous and strong. The establishment of our socialist system has opened the road leading to the ideal society of the future, but to translate this ideal into reality needs hard work. Some of our young people think that everything ought to be perfect once a socialist society is established and that they should be able to enjoy a happy life ready-made, without working for it. This is unrealistic.

VI. THE QUESTION OF THE MINORITY NATIONALITIES

The minority nationalities in our country number more than thirty million. Although they constitute only 6 per cent of the total population, they inhabit extensive regions which comprise 50 to 60 per cent of China's total area. It is thus imperative to foster good relations between the Han[9] people and the minority nationalities. The key to this question lies in overcoming Han chauvinism. At the same time, efforts should also be made to overcome local-nationality chauvinism,

[8] *Youth League:* the organization to train future CCP members. It was very important under Mao as the only legal nationwide organization besides the CCP and the People's Liberation Army (PLA).
[9] *Han:* Chinese name for the dominant ethnic group in China. The Han make up about 96 percent of China's population. The remaining citizens of the People's Republic of China (PRC) come from other ethnic groups, or nationalities, such as the Mongols, Manchus, or Uighurs, or from one of dozens of tribes.

wherever it exists among the minority nationalities. Both Han chauvinism and local-nationality chauvinism are harmful to the unity of the nationalities; they represent one kind of contradiction among the people which should be resolved. We have already done some work to this end. In most of the areas inhabited by minority nationalities, there has been considerable improvement in the relations between the nationalities, but a number of problems remain to be solved. In some areas, both Han chauvinism and local-nationality chauvinism still exist to a serious degree, and this demands full attention. As a result of the efforts of the people of all nationalities over the last few years, democratic reforms and socialist transformation have in the main been completed in most of the minority nationality areas. Democratic reforms have not yet been carried out in Tibet because conditions are not ripe. According to the seventeen-article agreement reached between the Central People's Government and the local government of Tibet, the reform of the social system must be carried out, but the timing can only be decided when the great majority of the people of Tibet and the local leading public figures consider it opportune, and one should not be impatient. It has now been decided not to proceed with democratic reforms in Tibet during the period of the Second Five-Year Plan.[10] Whether to proceed with them in the period of the Third Five-Year Plan can only be decided in the light of the situation at the time. . . .

[Section VII provides a one-page general reflection on "Overall consideration and proper arrangements."]

VIII. ON "LET A HUNDRED FLOWERS BLOSSOM, LET A HUNDRED SCHOOLS OF THOUGHT CONTEND" AND "LONG-TERM COEXISTENCE AND MUTUAL SUPERVISION"

"Let a hundred flowers blossom, let a hundred schools of thought contend" and "long-term coexistence and mutual supervision"—how did these slogans come to be put forward? They were put forward in the light of China's specific conditions, in recognition of the continued

[10]The First Five-Year Plan (1953–57) was modeled on the central economic planning of the Soviet Union. Such a planned economy replaced the market economy as the mechanism for determining investment, production, and spending. The Second Five-Year Plan ran from 1958 to 1962. China has followed such plans ever since, but their relevance has decreased with market reform in the post-Mao period.

existence of various kinds of contradictions in socialist society and in response to the country's urgent need to speed up its economic and cultural development. Letting a hundred flowers blossom and a hundred schools of thought contend is the policy for promoting progress in the arts and sciences and a flourishing socialist culture in our land. Different forms and styles in art should develop freely and different schools in science should contend freely. We think that it is harmful to the growth of art and science if administrative measures are used to impose one particular style of art or school of thought and to ban another. Questions of right and wrong in the arts and sciences should be settled through free discussion in artistic and scientific circles and through practical work in these fields. They should not be settled in an over-simple manner. A period of trial is often needed to determine whether something is right or wrong. Throughout history, at the outset new and correct things often failed to win recognition from the majority of people and had to develop by twists and turns through struggle. Often, correct and good things were first regarded not as fragrant flowers but as poisonous weeds. Copernicus' theory of the solar system and Darwin's theory of evolution were once dismissed as erroneous and had to win out over bitter opposition. Chinese history offers many similar examples. In a socialist society, the conditions for the growth of the new are radically different from and far superior to those in the old society. Nevertheless, it often happens that new, rising forces are held back and sound ideas stifled. Besides, even in the absence of their deliberate suppression, the growth of new things may be hindered simply through lack of discernment. It is therefore necessary to be careful about questions of right and wrong in the arts and sciences, to encourage free discussion and avoid hasty conclusions. We believe that such an attitude will help ensure a relatively smooth development of the arts and sciences.

Marxism, too, has developed through struggle. At the beginning, Marxism was subjected to all kinds of attack and regarded as a poisonous weed. This is still the case in many parts of the world. In the socialist countries, it enjoys a different position. But non-Marxist and, what is more, anti-Marxist ideologies exist even in these countries. In China, although socialist transformation has in the main been completed as regards the system of ownership, and although the large-scale, turbulent class struggles of the masses characteristic of times of revolution have in the main come to an end, there are still remnants of the overthrown landlord and comprador[11] classes, there is still a bourgeoisie, and the remoulding of the petty bourgeoisie has only just started. Class struggle is by no means over. The class struggle between the proletariat and the bourgeoisie, the class struggle between the various political forces, and the class struggle between the proletariat and the bourgeoisie in the ideological field will still be protracted and tortuous and at times even very sharp. The proletariat seeks to transform the world according to its own world outlook, and so does the bourgeoisie. In this respect, the question of which will win out, socialism or capitalism, is not really settled yet. Marxists remain a minority among the entire population as well as among the intellectuals. Therefore, Marxism must continue to develop through struggle. Marxism can develop only through struggle, and this is not only true of the past and the present, it is necessarily true of the future as well. What is correct invariably develops in the course of struggle with what is wrong. The true, the good and the beautiful always exist by contrast with the false, the evil and the ugly, and grow in struggle with them. As soon as something erroneous is rejected and a particular truth accepted by mankind, new truths begin to struggle with new errors. Such struggles will never end. This is the law of development of truth and, naturally, of Marxism.

It will take a fairly long period of time to decide the issue in the ideological struggle between socialism and capitalism in our country. The reason is that the influence of the bourgeoisie and of the intellectuals who come from the old society, the very influence which constitutes their class ideology, will persist in our country for a long time. If this is not understood at all or is insufficiently understood, the gravest of mistakes will be made and the necessity of waging struggle in the ideological field will be ignored. Ideological struggle differs from other forms of struggle, since the only method used is painstaking reasoning, and not crude coercion. Today, socialism is in an advantageous position in the ideological struggle. The basic power of the state is in the hands of the working people led by the proletariat. The Communist Party is strong and its prestige high. Although there are defects and mistakes in our work, every fair-minded person can see that we are loyal to the people, that we are both determined and able to build up our motherland together with them, and that we have already

[11]*comprador:* a nineteenth-century Chinese who worked with European merchants as a middleman in one of the various ports of China. Compradors were widely held to be unpatriotic, if not outright traitors. Many became wealthy and were among the new class of industrial leaders.

achieved great successes and will achieve still greater ones. The vast majority of the bourgeoisie and the intellectuals who come from the old society are patriotic and are willing to serve their flourishing socialist motherland; they know they will have nothing to fall back on and their future cannot possibly be bright if they turn away from the socialist cause and from the working people led by the Communist Party.

People may ask, since Marxism is accepted as the guiding ideology by the majority of the people in our country, can it be criticized? Certainly it can. Marxism is scientific truth and fears no criticism. If it did, and if it could be overthrown by criticism, it would be worthless. In fact, aren't the idealists criticizing Marxism every day and in every way? And those who harbour bourgeois and petty-bourgeois ideas and do not wish to change—aren't they also criticizing Marxism in every way? Marxists should not be afraid of criticism from any quarter. Quite the contrary, they need to temper and develop themselves and win new positions in the teeth of criticism and in the storm and stress of struggle. Fighting against wrong ideas is like being vaccinated—a man develops greater immunity from disease as a result of vaccination. Plants raised in hothouses are unlikely to be hardy. Carrying out the policy of letting a hundred flowers blossom and a hundred schools of thought contend will not weaken, but strengthen, the leading position of Marxism in the ideological field.

What should our policy be towards non-Marxist ideas? As far as unmistakable counter-revolutionaries and saboteurs of the socialist cause are concerned, the matter is easy, we simply deprive them of their freedom of speech. But incorrect ideas among the people are quite a different matter. Will it do to ban such ideas and deny them any opportunity for expression? Certainly not. It is not only futile but very harmful to use crude methods in dealing with ideological questions among the people, with questions about man's mental world. You may ban the expression of wrong ideas, but the ideas will still be there. On the other hand, if correct ideas are pampered in hothouses and never exposed to the elements and immunized against disease, they will not win out against erroneous ones. Therefore, it is only by employing the method of discussion, criticism and reasoning that we can really foster correct ideas and overcome wrong ones, and that we can really settle issues.

It is inevitable that the bourgeoisie and petty bourgeoisie will give expression to their own ideologies. It is inevitable that they will stubbornly assert themselves on political and ideological questions by every possible means. You cannot expect them to do otherwise. We should not use the method of suppression and prevent them from expressing themselves, but should allow them to do so and at the same time argue with them and direct appropriate criticism at them. Undoubtedly, we must criticize wrong ideas of every description. It certainly would not be right to refrain from criticism, look on while wrong ideas spread unchecked and allow them to dominate the field. Mistakes must be criticized and poisonous weeds fought wherever they crop up. However, such criticism should not be dogmatic, and the metaphysical method should not be used, but instead the effort should be made to apply the dialectical method. What is needed is scientific analysis and convincing argument. Dogmatic criticism settles nothing. We are against poisonous weeds of whatever kind, but we must carefully distinguish between what is really a poisonous weed and what is really a fragrant flower. Together with the masses of the people, we must learn to differentiate carefully between the two and use correct methods to fight the poisonous weeds.

At the same time as we criticize dogmatism, we must direct our attention to criticizing revisionism. Revisionism, or Right opportunism, is a bourgeois trend of thought that is even more dangerous than dogmatism. The revisionists, the Right opportunists, pay lip-service to Marxism; they too attack "dogmatism." But what they are really attacking is the quintessence of Marxism. They oppose or distort materialism and dialectics, oppose or try to weaken the people's democratic dictatorship and the leading role of the Communist Party, and oppose or try to weaken socialist transformation and socialist construction. Even after the basic victory of our socialist revolution, there will still be a number of people in our society who vainly hope to restore the capitalist system and are sure to fight the working class on every front, including the ideological one. And their right-hand men in this struggle are the revisionists.

Literally the two slogans—let a hundred flowers blossom and let a hundred schools of thought contend—have no class character; the proletariat can turn them to account, and so can the bourgeoisie or others. Different classes, strata and social groups each have their own views on what are fragrant flowers and what are poisonous weeds. Then, from the point of view of the masses, what should be the criteria today for distinguishing fragrant flowers from poisonous weeds? In their political activities, how should our people judge whether a person's words and deeds are right or wrong? On the basis of the principles of our Constitution, the will of the overwhelming majority of our people and the common political positions which have been proclaimed on various occasions by our political parties, we consider that, broadly speaking, the criteria should be as follows:

1. Words and deeds should help to unite, and not divide, the people of all our nationalities.
2. They should be beneficial, and not harmful, to socialist transformation and socialist construction.
3. They should help to consolidate, and not undermine or weaken, the people's democratic dictatorship.
4. They should help to consolidate, and not undermine or weaken, democratic centralism.
5. They should help to strengthen, and not shake off or weaken, the leadership of the Communist Party.
6. They should be beneficial, and not harmful, to international socialist unity and the unity of the peace-loving people of the world.

Of these six criteria, the most important are the two about the socialist path and the leadership of the Party. These criteria are put forward not to hinder but to foster the free discussion of questions among the people. Those who disapprove these criteria can still state their own views and argue their case. However, so long as the majority of the people have clear-cut criteria to go by, criticism and self-criticism can be conducted along proper lines, and these criteria can be applied to people's words and deeds to determine whether they are right or wrong, whether they are fragrant flowers or poisonous weeds. These are political criteria. Naturally, to judge the validity of scientific theories or assess the aesthetic value of works of art, other relevant criteria are needed. But these six political criteria are applicable to all activities in the arts and sciences. In a socialist country like ours, can there possibly be any useful scientific or artistic activity which runs counter to these political criteria?

The views set out above are based on China's specific historical conditions. Conditions vary in different socialist countries and with different Communist Parties. Therefore, we do not maintain that they should or must adopt the Chinese way.

The slogan "long-term coexistence and mutual supervision" is also a product of China's specific historical conditions. It was not put forward all of a sudden, but had been in the making for several years. The idea of long-term coexistence had been there for a long time. When the socialist system was in the main established last year, the slogan was formulated in explicit terms. Why should the bourgeois and petty-bourgeois democratic parties be allowed to exist side by side with the party of the working class over a long period of time? Because we have no reason for not adopting the policy of long-term coexistence with all those political parties which are truly devoted to the task of uniting the people for the cause of socialism and which enjoy the trust of the people. .... It is the desire as well as the policy of the Communist Party to exist side by side with the democratic parties for a long time to come. But whether the democratic parties can long remain in existence depends not merely on the desire of the Communist Party but on how well they acquit themselves and on whether they enjoy the trust of the people. Mutual supervision among the various parties is also a long-established fact, in the sense that they have long been advising and criticizing each other. Mutual supervision is obviously not a one-sided matter; it means that the Communist Party can exercise supervision over the democratic parties, and vice versa. Why should the democratic parties be allowed to exercise supervision over the Communist Party? Because a party as much as an individual has great need to hear opinions different from its own. We all know that supervision over the Communist Party is mainly exercised by the working people and the Party membership. But it augments the benefit to us to have supervision by the democratic parties too. Of course, the advice and criticism exchanged by the Communist Party and the democratic parties will play a positive supervisory role only when they conform to the six political criteria given above. Thus, we hope that in order to fit in with the needs of the new society, all the democratic parties will pay attention to ideological remoulding and strive for long-term coexistence with the Communist Party and mutual supervision.

IX. ON THE QUESTION OF DISTURBANCES CREATED BY SMALL NUMBERS OF PEOPLE

In 1956, small numbers of workers or students in certain places went on strike. The immediate cause of these disturbances was the failure to satisfy some of their demands for material benefits, of which some should and could have been met, while others were out of place or excessive and therefore could not be met for the time being. But a more important cause was bureaucracy[12] on the part of the leadership. In some cases, the responsibility for such bureaucratic mistakes fell

---

[12]*bureaucracy:* the excessive and oppressive use of bureaucratic power by government officials. The term Mao used here, *guanliaozhuyi,* is more often translated as "bureaucratism."

on the higher authorities, and those at the lower levels were not to blame. Another cause of these disturbances was lack of ideological and political education among the workers and students. The same year, in some agricultural co-operatives there were also disturbances created by a few of their members, and here too the main causes were bureaucracy on the part of the leadership and lack of educational work among the masses.

It should be admitted that among the masses some are prone to pay attention to immediate, partial and personal interests and do not understand, or do not sufficiently understand, long-range, national and collective interests. Because of lack of political and social experience, quite a number of young people cannot readily see the contrast between the old China and the new, and it is not easy for them thoroughly to comprehend the hardships our people went through in the struggle to free themselves from the oppression of the imperialists and Guomindang reactionaries, or the long years of hard work needed before a fine socialist society can be established. That is why we must constantly carry on lively and effective political education among the masses and should always tell them the truth about the difficulties that crop up and discuss with them how to surmount these difficulties.

We do not approve of disturbances, because contradictions among the people can be resolved through the method of "unity—criticism—unity," while disturbances are bound to cause some losses and are not conducive to the advance of socialism. We believe that the masses of the people support socialism, conscientiously observe discipline and are reasonable, and will certainly not take part in disturbances without cause. But this does not mean that the possibility of disturbances by the masses no longer exists in our country. On this question, we should pay attention to the following. (1) In order to root out the causes of disturbances, we must resolutely overcome bureaucracy, greatly improve ideological and political education, and deal with all contradictions properly. If this is done, generally speaking there will be no disturbances. (2) When disturbances do occur as a result of poor work on our part, then we should guide those involved onto the correct path, use the disturbances as a special means for improving our work and educating the cadres and the masses, and find solutions to those problems which were previously left unsolved. In handling any disturbance, we should take pains and not use over-simple methods, or hastily declare the matter closed. The ringleaders in disturbances should not be summarily expelled, except for those who have committed criminal offences or are active counter-revolutionaries and have to be punished by law. In a large country like ours, there is nothing to get alarmed about if small numbers of people create disturbances; on the contrary, such disturbances will help us get rid of bureaucracy.

There are also a small number of individuals in our society who, flouting the public interest, wilfully break the law and commit crimes. They are apt to take advantage of our policies and distort them, and deliberately put forward unreasonable demands in order to incite the masses, or deliberately spread rumours to create trouble and disrupt public order. We do not propose to let these individuals have their way. On the contrary, proper legal action must be taken against them. Punishing them is the demand of the masses, and it would run counter to the popular will if they were not punished.

X. CAN BAD THINGS BE TURNED INTO GOOD THINGS?

In our society, as I have said, disturbances by the masses are bad, and we do not approve of them. But when disturbances do occur, they enable us to learn lessons, to overcome bureaucracy and to educate the cadres and the masses. In this sense, bad things can be turned into good things. Disturbances thus have a dual character. Every disturbance can be regarded in this way.

Everybody knows that the Hungarian incident was not a good thing. But it too had a dual character. Because our Hungarian comrades took proper action in the course of the incident, what was a bad thing has eventually turned into a good one. Hungary is now more consolidated than ever, and all other countries in the socialist camp have also learned a lesson. . . .

People all over the world are now discussing whether or not a third world war will break out. On this question, too, we must be mentally prepared and do some analysis. We stand firmly for peace and against war. But if the imperialists insist on unleashing another war, we should not be afraid of it. Our attitude on this question is the same as our attitude towards any disturbance: first, we are against it; second, we are not afraid of it. The First World War was followed by the birth of the Soviet Union with a population of 200 million. The Second World War was followed by the emergence of the socialist camp with a combined population of 900 million. If the imperialists insist on launching a third world war, it is certain that several hundred million more will turn to socialism, and then there will not be much room left on earth for the imperialists; it is also likely that the whole structure of imperialism will completely collapse.

In given conditions, each of the two opposing aspects of a contradiction invariably transforms itself into its opposite as a result of the struggle between them. Here, it is the conditions which are essential. Without the given conditions, neither of the two contradictory aspects can transform itself into its opposite. Of all the classes in the world the proletariat is the one which is most eager to change its position, and next comes the semi-proletariat, for the former possesses nothing at all while the latter is hardly any better off. The United States now controls a majority in the United Nations and dominates many parts of the world—this state of affairs is temporary and will be changed one of these days. China's position as a poor country denied its rights in international affairs will also be changed—the poor country will change into a rich one, the country denied its rights into one enjoying them—a transformation of things into their opposites. Here, the decisive conditions are the socialist system and the concerted efforts of a united people. . . .

[In Section XI, Mao uses a page and a half to urge the party to economize and "share weal and woe with the masses."]

XII. CHINA'S PATH TO INDUSTRIALIZATION

In discussing our path to industrialization, we are here concerned principally with the relationship between the growth of heavy industry, light industry and agriculture. It must be affirmed that heavy industry is the core of China's economic construction. At the same time, full attention must be paid to the development of agriculture and light industry. . . .

I do not propose to dwell on economic questions today. With barely seven years of economic construction behind us, we still lack experience and need to accumulate it. Neither had we any experience in revolution when we first started, and it was only after we had taken a number of tumbles and acquired experience that we won nation-wide victory. What we must now demand of ourselves is to gain experience in economic construction in a shorter period of time than it took us to gain experience in revolution, and not to pay as high a price for it. Some price we will have to pay, but we hope it will not be as high as that paid during the period of revolution. We must realize that there is a contradiction here—the contradiction between the objective laws of economic development of a socialist society and our subjective cognition of them—which needs to be resolved in the course of practice. This contradiction also manifests itself as a contradiction between different people, that is, a contradiction between those in whom the reflection of these objective laws is relatively accurate and those in whom the reflection is relatively inaccurate; this, too, is a contradiction among the people. Every contradiction is an objective reality, and it is our task to reflect it and resolve it in as nearly correct a fashion as we can.

In order to turn China into an industrial country, we must learn conscientiously from the advanced experience of the Soviet Union. The Soviet Union has been building socialism for forty years, and its experience is very valuable to us. Let us ask: Who designed and equipped so many important factories for us? Was it the United States? Or Britain? No, neither the one nor the other. Only the Soviet Union was willing to do so, because it is a socialist country and our ally. In addition to the Soviet Union, the fraternal countries in East Europe have also given us some assistance. It is perfectly true that we should learn from the good experience of all countries, socialist or capitalist, about this there is no argument. But the main thing is still to learn from the Soviet Union. Now there are two different attitudes towards learning from others. One is the dogmatic attitude of transplanting everything, whether or not it is suited to our conditions. This is no good. The other attitude is to use our heads and learn those things which suit our conditions, that is, to absorb whatever experience is useful to us. That is the attitude we should adopt.

To strengthen our solidarity with the Soviet Union, to strengthen our solidarity with all the socialist countries—this is our fundamental policy, this is where our basic interests lie. Then there are the Asian and African countries and all the peace-loving countries and peoples— we must strengthen and develop our solidarity with them. United with these two forces, we shall not stand alone. As for the imperialist countries, we should unite with their people and strive to coexist peacefully with those countries, do business with them and prevent a possible war, but under no circumstances should we harbour any unrealistic notions about them.

8

Talks at the Beidaihe Conference

August 1958

Mao's talks at the Beidaihe conference of the CCP politburo have never been officially published in China. But scholars have long known that this meeting in August 1958 produced the big push for Mao's utopian scheme, the "people's communes," and ushered in the high tide of the Great Leap Forward. This text contains extracts from a "draft transcript" of Mao's five long, rambling talks given between August 17 and 30 at the beachside party retreat. The transcript appears in a Cultural Revolution–period "genius" Mao edition that came to light in the 1980s (see "A Note about the Texts"). The extracts here are taken in chronological order from the forty-five-page text, with an emphasis on Mao's comments on the people's communes.

Mao's utopian visions were at their height at the Beidaihe Conference. His wild romanticizing of the Yan'an experience and his equating public dining halls and a nonmonetary economy with communism are startling aspects of the speeches. This shows how little concern Mao had for proper Marxist terms. His colleagues were perplexed by the blending of his noble ideas (such as the dignity of manual labor and a better deal for women and youth) with his unrealistic musings on "deep ploughing" and "airports" for every commune. This selection also includes his tragic call to increase China's population. Two long-term results of Mao's utopian visions have been noted: The improvement of women's place in China has been needlessly linked to the excesses of the Cultural Revolution (and neatly ignored in post-Mao China), and China now wrestles with a severe overpopulation problem.

---

17 AUGUST

This is an enlarged Politburo conference; responsible comrades from all provinces and autonomous regions are participating. The topics are

---

Xuexi wenxuan (Study Selections) (n.p., n.d.), 295–321. Translation from Roderick Mac-Farquhar, Timothy Cheek, and Eugene Wu, eds., *The Secret Speeches of Chairman Mao: From the Hundred Flowers to the Great Leap Forward* (Cambridge: Harvard Council on East Asian Studies, 1989), 397–400, 403, 408, 418–19, 429–30, 434–36, 441.

listed in the documents distributed here, and comrades may refer to them.

The key point is the first problem: the problem of the Five Year Economic Plan next year—mainly it has to do with industry, but it's also somewhat relevant to agriculture. It's unfair just to issue reference figures; we should be fairer and more correct with our figures. Let's take three days to work on it; Comrade Li Fuchun* will be responsible....

The eighth problem: the problem of cadres participating in manual labor. Officials, no matter who, whether big or small, including us here, all should participate in manual labor as long as they're physically able, excepting only those too old or too weak. We have millions of officials; adding those in the army there are more than ten million. We don't even have a clear idea as to exactly how many officials. Cadre children number in the tens of millions, and they are in a favored position from which to become officials. When one has been an official for a long time, it's easy to get separated from reality and the masses. The construction of the Ming Tombs Reservoir has been completed; many people went to the reservoir to perform manual labor for a few days.† Can we do manual labor for one month a year, making assignments according to the four seasons of the year? Workers, peasants, and commercial personnel are all able to combine manual labor and their ordinary work; everybody should be like that. If other people can do manual labor, is it acceptable that our officials should not? And then there are all those many cadre children. In the Soviet Union, graduates of agricultural colleges don't want to go to the countryside. Isn't it absurd to run agricultural colleges in the cities! Agricultural colleges should all move to the countryside. All schools should run factories. Even the Tianjin Conservatory is running a few factories; that's very good. Participating in manual labor is easy for cadres at the county and township levels, but it's hard to manage for those at the level of the Center, provinces, and prefectures; they probably can't operate a machine! How is it that people who can eat with chopsticks and write with brushes can't operate a machine? Is it easier to operate a machine or to climb a mountain? ...

The fifteenth problem: the problem of deep ploughing. Currently the main orientation in agriculture is the problem of deep ploughing.

*_Li Fuchun:_ Chairman, State Planning Commission
†Mao himself had led his senior colleagues to work there in May to symbolize the importance he placed on manual labor.

Ploughing deeply is like creating a big reservoir for water and a big cistern for manure, otherwise no amount of water and manure will work. In the north we should deep plough to a depth of over a foot, in the south to 7 or 8 inches, then apply manure in different layers so as to enhance the granular structure of the soil. Every granule is both a small reservoir, and small manure cistern. Deep ploughing brings water above ground into contact with underground water. Close planting is based on deep ploughing: otherwise it is useless. Deep ploughing helps weeding. Digging up roots in turn helps eliminate insects so that one *mou*[1] of land can produce as much as three *mou*. Now each person nationwide has three *mou* of land. Once we go down to the grass roots, we can increase production. What's the use of planting that much land? In the future we can use one-third of the land for afforestation, and after a few years again decrease the acreage for grain by another *mou* per capita. In the past we weren't able to afforest the plains, but by that time we will be able to. Without deep ploughing there's no such possibility.

Our views on population should change. In the past I said that we could manage with 800 million. Now I think that one billion plus would be no cause for alarm. This shouldn't be recommended for people with many children. When people's level of education increases, they will really practice birth control. . . .

Now on the problem of the people's communes: What should they be called? They may be called people's communes, or they may not. My inclination is to call them people's communes. This name is still socialist in nature, not at all overemphasizing communism. They're called people's communes, first, because they're big and, second because they're public. Lots of people, a vast area of land, large scale of production, and all their undertakings are done in a big way. They integrate government administration with commune management to establish public mess halls, and private plots are eliminated. But chickens, ducks, and the young trees in front and behind a house are still private. These, of course, won't exist in the future. If we have more grain, we can practice the supply system; for the present it's still reward according to one's work. Wages will be given to individuals according to their ability and won't be given to the head of the family, which makes the youth and women happy. This will be very beneficial for the liberation of the individual. In establishing the people's communes, as I see it, once again it's been the countryside that's taken

[1]mou: *mu*

the lead; the cities haven't started yet, because the workers' wage scales are a complicated matter. Whether in urban or rural areas, the aim should be the socialist system plus communist ideology. The Soviet Union practices the use of high rewards and heavy punishments, emphasizing only material incentives. We now practice socialism, and have the sprouts of communism. Schools, factories, and neighborhoods can all establish people's communes. In a few years big communes will be organized to include everyone. . . .

The people's communes contain the sprouts of communism. When products are bountiful, we will implement communist distribution of grain, cotton, and edible oils. By that time morality will have greatly improved. Labor will no longer require supervision; even if you want someone to rest, he won't. The Jianhua Machinery Plant practices the *"ba wu."** In the people's communes, people practice cooperation in a big way, bringing their own tools and food. The workers beat drums and gongs and don't ask for piece-rate wages. All of these are the sprouts of communism, and they destroy the system of bourgeois right. I hope everyone will publicize this way of looking at the problem, read the two relevant documents, and publicize the actual situations in which there is increasing growth of the elements of communist morality.

In the past during the revolution numerous people died without asking anything in return. Why can't it be like that now? If we can eat without paying for it, this will be a tremendous change. Probably in about ten years our production will be very bountiful and the people's morality will be very noble; then we can practice communism in eating, clothing, and housing. Eating without paying in public mess halls is communism. In the future everything will be called a commune. We won't say factories; for instance, Anshan Steel Mill will be called Anshan Commune. Cities and villages will be called communes, and universities and neighborhoods will establish communes. Townships will be integrated with communes, local government will be integrated with commune management; temporarily we put up two signs. Set up a "department of the interior" (administrative section) in the people's communes to administer registry of birth and death, marriage, census, and civil administration. . . . The characteristics of the people's

---

*Ba wu: literally, eight have-nots. This term is obscure, but one Chinese respondent who was a high school student during the Great Leap Forward partially recalls learning a chant, "The Eight Proletarian Points" *(bage wuchanjieji zhuzhang)*. . . . *Ba wu* would not be an unusual contraction, just as *si hua* is the contraction for *sige xiandai hua* (the four modernizations slogan).

communes are one, big, and two, public; most important is that many cooperatives combine into one big commune. The several comments in *Socialist Upsurge in the Countryside** said big cooperatives were good; mountain areas could establish big cooperatives, too, in order to develop a diversified economy and to ensure all-round development. However, establishing slightly small cooperatives to begin with also had its advantages. The youth and women are happy about the new wage system. The rationale of increasing private plots and so on were all proposed by the party's Rural Work Department.† As early as 1955 I recommended establishing big cooperatives. Establish 15,000 to 25,000 communes nationwide averaging 5,000 to 6,000 households or 20,000 to 30,000 people per commune—rather large and convenient for running industry, agriculture, commerce, education, and military affairs side by side, as well as farming, forestry, animal husbandry, sideline production and fisheries. With this way of doing things, I think in the future a few large cities will be dispersed; residential areas of 20,000 to 30,000 people will have everything; villages will become small cities where the majority of philosophers and scientists will be assigned. Every large commune will have highways constructed, wider roads of cement or asphalt, with no trees planted alongside so that airplanes can land—they will be airports. In the future, every province should have a couple of hundred airplanes, averaging two planes per township. Large provinces will set up their own factories for aircraft construction‡ . . .

30 AUGUST, MORNING

The first precondition for communism is plenty; the second is to have a communist spirit. Once an order is issued, everyone automatically goes to their work, idlers are few or none. Communism does not differentiate between superiors and subordinates. We have a twenty-two year history of war communism, with no salaries, which is different from the Soviet Union. In the Soviet Union it's called the system of

---

*The full title of this collection of essays on collectivization edited by Mao and published in January 1956 was *Socialist Upsurge in China's Countryside (Zhongguo nongcun de shehui-zhuyi gaochao)*. Mao seems to be concerned here to demonstrate the consistency between his views of late 1955, when he completed his editorial notes to this book, and August 1958.

†This is a dig at the then head of the department, Deng Zihui, who was criticized in 1955 for being too cautious on collectivization.

‡This utopian vision seems redolent of Sun Yat-sen's earlier visions of the industrialization of China; see C. Martin Wilbur, *Sun Yat-sen, Frustrated Patriot* (New York: Columbia University Press, 1976), 23–26.

surplus grain collection. We didn't practice that. Ours was called the supply system, in which army and civilians, officers and men are equal, and there are also the three great democracies.* Originally we divided up the leftovers from the mess and had small subsidies. After we came into the cities, it was said that the supply system was backward, guerrillaism, a rural work style, and that it couldn't boost initiative, nor stimulate progress. They wanted to establish a salary system. They endured for three years, and in 1952 the salary system was established. They said bourgeois ranks and rights and such were very fine and called our old supply system a backward method, a guerrilla practice that affected activism. In effect they turned the supply system into a system of bourgeois right, thereby promoting bourgeois ideology. Did initiating the 25,000 *li* Long March, the Land Revolution and the War of Liberation rely on salaries? Two to three million people during the anti-Japanese war, from four to five million during the War of Liberation lived a life of war communism, no Sundays off—didn't they all risk their lives? The party, the administration, the army, the civilians—numbering several million—all were together with the masses, supporting the administration and loving the people. The party, the administration, and the army under a unified leadership had nothing "to spend," but with unity between officers and men, and between the army and civilians, and the support for the administration and love for the people, we drove off the Japanese devils and defeated Chiang Kai-shek. Nor did we have anything "to spend" when we fought the United States. Can it be said we did all this because we handed out salaries? Now we have something "to spend," issuing salaries according to rank, dividing them into generals, field rank officers, and junior officers; but some of them have not even been in battle. Whether or not they're any good has yet to be tested. The result is divorce from the masses; the men don't love their officers, and the masses don't love their cadres. Because of this we're not much different from the Nationalist party: our garments are in three colors, our food is divided into five grades, even the desks and chairs of our offices are ranked; and so the workers and peasants don't like us, saying, "You're officials—party officials, government officials, military officials, commercial officials"—so many officials, how can there be no "isms"?† Too many bureaucratic airs, too little politics, so

---

*Democracy in military affairs, politics, and economics.

†Mao is presumably referring to the three sins combatted in the previous year's Rectification Campaign: bureaucratism, sectarianism, and subjectivism.

bureaucratism emerges. Since the Rectification Campaign we have been rectifying bureaucratic airs and putting politics in command. Since then the cases of competing for rank and scrambling for special treatment have not been many. I think we should get rid of this thing. The salary system does not have to be abolished immediately, because there are professors. But we should prepare for it in one or two years. Once the people's communes are established, this will force us gradually to abolish the salary system. Since we came into the cities, we have been under the influence of the bourgeoisie. When we launched a campaign, it was a really Marxist practice and a democratic work style, but they branded us as using "rural work style" and "guerrilla practices." "Guerrilla practices" are capitalists' words. It was probably during the period from 1953 to mid 1957* when they did things together with the bourgeoisie, local tyrants, and evil gentry that they began to straighten their clothes, sit properly, and study the bourgeois style—having haircuts and shaves, shaving three times a day— all learned from the same source. . . .

On eliminating the four pests:† We can focus on this during the National Day holidays, the New Year, and the Lunar New Year. The less we have of these four pests, the better for us; because these four pests harm the people and directly affect the health of the people. We must wipe out various kinds of diseases on a large scale. At one place in Hangzhou, only one person took sick last year. The rate of attendance at work exceeded 90 percent. When the physicians have nothing to do, they can go till the fields and do research. The day China eliminates the four pests we should hold a celebration. It should be recorded in history books. The capitalist states have not done it; those so-called civilized states still have many flies and mosquitoes.

*In mid 1957 the bourgeoisie was counterattacked in the Anti-Rightist Campaign.
†four pests: rats, sparrows, flies, mosquitoes. In 1960, bed bugs were substituted for sparrows when it was realized that the extermination of these birds had permitted insect pests to flourish.

9

American Imperialism Is Closely Surrounded by the Peoples of the World

1964

China and the United States were enemies in the cold war. Since the Communist victory in China in 1949, the United States enforced an embargo on the People's Republic of China. The 1960s were a time of increasing tension, as the United States became involved in a land war in Southeast Asia, as well as various adventures in Africa. In this short message of support to the people of the Congo, Mao reflects his and the Chinese government's view of America as the world's biggest imperialist aggressor. Yet Mao draws on Chinese revolutionary experience to claim that protracted resistance by people in any area of the world will ulti-mately prevail over American imperialism. In hindsight we can see that Mao was more accurate in his hopes for the Vietnamese than the African fighters, but his call still rings around the world for groups that take up arms against what they see as a U.S.-dominated international economic order (such as the Shining Path in Peru and Marxist rebels in Nepal). In this message, Mao also makes one version of his famous claim that U.S. strength is a "paper tiger." The virulent anti-Americanism of this piece is representative of the rhetoric of the 1960s (paralleled by Ameri-can denunciations of "Red China"). Thus it was stunning when a few years later, in the early 1970s, Mao and U.S. president Richard Nixon opened trade relations and joined forces against the Soviet Union.

The U.S. imperialist armed aggression against the Congo (Leopold-ville) is a very grave matter.

The United States has all along attempted to control the Congo. It used the United Nations forces to carry out every sort of evil deed there. It murdered the Congolese national hero Lumumba, it sub-verted the lawful Congolese government. It imposed the puppet Tshombe on the Congolese people, and dispatched mercenary troops

Originally translated in *Peking Review.* Adaptation from Stuart Schram, *The Political Thought of Mao Tse-tung,* rev. ed. (New York: Praeger, 1969), 383–84.

to suppress the Congolese national liberation movement. And now, it is carrying out direct armed intervention in the Congo in collusion with Belgium and Britain. In so doing, the purpose of U.S. imperialism is not only to control the Congo, but also to enmesh the whole of Africa, particularly the newly independent African countries, in the toils of U.S. neo-colonialism once again. U.S. aggression has encountered heroic resistance from the Congolese people and aroused the indignation of the people of Africa and of the whole world.

U.S. imperialism is the common enemy of the people of the whole world. It is engaged in aggression against South Vietnam, it is intervening in Laos, menacing Cambodia and blustering about extending the war in Indochina. It is trying everything to strangle the Cuban revolution. It wants to turn West Germany and Japan into two important nuclear bases of the United States. It ganged up with England in creating so-called Malaysia to menace Indonesia and other south-east Asian countries. It is occupying South Korea and China's Taiwan province. It is dominating all Latin America. It rides roughshod everywhere. U.S. imperialism has over-extended its reach. It adds a new noose around its neck every time it commits aggression anywhere. It is closely surrounded by the people of the whole world.

In their just struggle, the Congolese people are not alone. All the Chinese people support you. All the people throughout the world who oppose imperialism support you. U.S. imperialism and the reactionaries of all countries are paper tigers. The struggle of the Chinese people proved this. The struggle of the Vietnamese people is now proving it. The struggle of the Congolese people will certainly prove it too. Strengthening national unity and persevering in protracted struggle, the Congolese people will certainly be victorious, and U.S. imperialism will certainly be defeated.

People of the world, unite and defeat the U.S. aggressors and all their running dogs! People of the world, be courageous, dare to fight, defy difficulties and advance wave upon wave. Then the whole world will belong to the people. Monsters of all kinds shall be destroyed.

10

Cultural Revolution Readings

1960s

The Cultural Revolution saw the apotheosis of Chairman Mao. Although party leaders were aware of the tensions between the Chairman and his closest colleagues over the best ways to achieve socialism, and especially over how to recover from the Great Leap Forward, the public saw only their revered, godlike leader—the "Great Helmsman," the "Savior of the Chinese People." For reasons that still defy simple explanations, many Chinese acted irrationally in the Cultural Revolution (1966–69).[1] *Senior leaders found it impossible to bring themselves to stop Mao in the beginning and were helpless to save themselves soon thereafter. "Bombard the Headquarters" is the title of Mao's August 5, 1966, "big-character poster" denouncing "some leading comrades" for arrogance and opposing the proletariat. Big-character posters are essays written on big sheets of paper and hung on the walls of factories, schools, or other places where people lived and worked. Anyone could write and post these essays. Mao's big-character poster galvanized the emerging Red Guard movement of college and high school students. Leaders, from politburo members to local schoolteachers, were paraded through the streets in the dunce caps that Hunan rebels had used on evil landlords in Mao's 1927 "Report on the Peasant Movement in Hunan" (see Document 1). The Cultural Revolution was under way.*

The bible for the Red Guards was Quotations from Chairman Mao Zedong, *more commonly known as the "Little Red Book." Hundreds of millions of copies were printed and distributed during the Cultural Revolution.* Quotations *was a collection of brief quotations from Mao's published works on thirty-three topics, ranging from "The Communist Party"*

---

[1]Today most people refer to the "Cultural Revolution decade" of 1966–76, as Mao's radical policies and the political infighting he instigated in 1966 dominated the decade. However, Mao declared the Cultural Revolution over—that is, "a success"—at the Ninth Congress in April 1969.

"Bombard the Headquarters," New China News Agency, 4 Aug. 1967; *Quotations from Chairman Mao Tse-tung* (Peking: Foreign Languages Press, 1968), i–iv, 1–3, 8–10, 294–98, 312; "Just a Few Words," in *"Zai Zhongyang gongzuo huiyi shang de jianghu"* (Talk at the Central Work Conference), 25 Oct. 1966, translation from Stuart R. Schram, ed., *Mao Tse-tung Unrehearsed: Talks and Letters 1956–1971* (Harmondsworth: Penguin, 1974), 270–74.

to *"People's War"* to *"Unity"* to *"Women"* to *"Study."* Lin Biao, the defense minister, had created it as a basic primer for People's Liberation Army soldiers in the early 1960s. In December 1966, Lin penned a new preface to a second edition of Quotations lauding Chairman Mao. He would become Mao's heir apparent in 1969, replacing Liu Shaoqi, who became the most prominent victim of the Cultural Revolution. Lin's effusive preface to Quotations is a perfect example of the doublespeak that came to dominate the Cultural Revolution. Faith had outstripped reason, and Lin's epigraph to Quotations sums up the mind-set well: *"Study Chairman Mao's writings, follow his teachings and act according to his instructions."*

Even though the quotations themselves carry citations to the original texts from which they are drawn, one cannot escape the fact that Mao's long essays have been reduced to "sound bites" taken out of context. Nonetheless, the section on women does reflect some of Mao's calls for the fairer treatment of women. Unfortunately, the repudiation of Mao's utopian policies in the post-Mao period has resulted in the neglect of many of his more sensible ideas on women's liberation.

Mao realized early on that the Cultural Revolution was getting out of hand. "Just a Few Words," his comments at a Central Party Work Conference on October 25, 1966, seems reasonable and self-critical, but as with his 1957 "Contradictions" speech (see Document 7), his apparent reasonableness was belied by his ruthless political maneuvering. By 1966 all CCP officials knew this, and his final words in this talk are apt: "Does anyone else want to speak? I guess that's all for today. The meeting is adjourned."

BOMBARD THE HEADQUARTERS, AUGUST 5, 1966

China's first Marxist-Leninist big-character poster and commentator's article on it in the *People's Daily*[2] are indeed superbly written! Comrades, please read them again. But in the last fifty days or so some leading comrades from the central down to the local levels have acted in a diametrically opposite way. Having the reactionary stand of the bourgeoisie, they have enforced a bourgeois dictatorship and struck

[2]Mao is referring to the big-character poster written by Nie Yuanzi at Beijing University in June 1966, which began the attacks on party leaders characteristic of the Cultural Revolution. Liu Shaoqi, president of China, and other leaders had limited this movement in June and July, but this August 5 big-character poster by Mao himself gave the new Red Guards permission to renew the attack.

down the surging movement of the Great Cultural Revolution of the Proletariat. They have stood facts on their head and juggled black and white, encircled and suppressed revolutionaries, stifled opinions differing from their own, imposed a white terror, and felt very pleased with themselves. They have puffed up the arrogance of the bourgeoisie and deflated the morale of the proletariat. How vicious they are! Viewed in connection with the right deviation in 1962 and the wrong tendency of 1964 which was "left" in form but right in essence, shouldn't this prompt one to deep thought?

QUOTATIONS FROM CHAIRMAN MAO ZEDONG, 1968

Study Chairman Mao's writings, follow his teachings and act according to his instructions.   —Lin Biao

Foreword to the Second Edition

Comrade Mao Zedong is the greatest Marxist-Leninist of our era. He has inherited, defended and developed Marxism-Leninism with genius, creatively and comprehensively and has brought it to a higher and completely new stage.

Mao Zedong's thought is Marxism-Leninism of the era in which imperialism is heading for total collapse and socialism is advancing to world-wide victory. It is a powerful ideological weapon for opposing imperialism and for opposing revisionism and dogmatism. Mao Zedong's thought is the guiding principle for all the work of the Party, the army and the country.

Therefore, the most fundamental task in our Party's political and ideological work is at all times to hold high the great red banner of Mao Zedong's thought, to arm the minds of the people throughout the country with it and to persist in using it to command every field of activity. The broad masses of the workers, peasants and soldiers and the broad ranks of the revolutionary cadres and the intellectuals should really master Mao Zedong's thought; they should all study Chairman Mao's writings, follow his teachings, act according to his instructions and be his good fighters.

In studying the works of Chairman Mao, one should have specific problems in mind, study and apply his works in a creative way, combine study with application, first study what must be urgently applied so as to get quick results, and strive hard to apply what one is studying. In order really to master Mao Zedong's thought, it is essential to study many of Chairman Mao's basic concepts over and over again, and it is best to memorize important statements and study and apply them repeatedly. The newspapers should regularly carry quotations from Chairman Mao relevant to current issues for readers to study and apply. The experience of the broad masses in their creative study and application of Chairman Mao's works in the last few years has proved that to study selected quotations from Chairman Mao with specific problems in mind is a good way to learn Mao Zedong's thought, a method conducive to quick results.

We have compiled *Quotations from Chairman Mao Zedong*[3] in order to help the broad masses learn Mao Zedong's thought more effectively. In organizing their study, units should select passages that are relevant to the situation, their tasks, the current thinking of their personnel, and the state of their work.

In our great motherland, a new era is emerging in which the workers, peasants and soldiers are grasping Marxism-Leninism, Mao Zedong's thought. Once Mao Zedong's thought is grasped by the broad masses, it becomes an inexhaustible source of strength and a spiritual atom bomb of infinite power. The large-scale publication of *Quotations from Chairman Mao Zedong* is a vital measure for enabling the broad masses to grasp Mao Zedong's thought and for promoting the revolutionization of our people's thinking. It is our hope that all comrades will learn earnestly and diligently, bring about a new nation-wide high tide in the creative study and application of Chairman Mao's works and, under the great red banner of Mao Zedong's thought, strive to build our country into a great socialist state with modern agriculture, modern industry, modern science and culture and modern national defence!

LIN BIAO

1. The Communist Party

The force at the core leading our cause forward is the Chinese Communist Party.

The theoretical basis guiding our thinking is Marxism-Leninism.
— *Opening address at the First Session of the First National People's Congress of the People's Republic of China (September 15, 1954).*

---

[3]In 1968, this official translation used the older romanization of Mao Zedong's name, Mao Tse-tung, but we have used the *pinyin* system here and throughout this text.

If there is to be revolution, there must be a revolutionary party. Without a revolutionary party, without a party built on the Marxist-Leninist revolutionary theory and in the Marxist-Leninist revolutionary style, it is impossible to lead the working class and the broad masses of the people in defeating imperialism and its running dogs.[4]
   —*"Revolutionary Forces of the World Unite, Fight Against Imperialist Aggression!" (November 1948), Selected Works, Vol. IV, p. 284.*

Without the efforts of the Chinese Communist Party, without the Chinese Communists as the mainstay of the Chinese people, China can never achieve independence and liberation, or industrialization and the modernization of her agriculture.
   —*"On Coalition Government" (April 24, 1945), Selected Works, Vol. III, p. 318.*

The Chinese Communist Party is the core of leadership of the whole Chinese people. Without this core, the cause of socialism cannot be victorious.
   —*Talk at the general reception for the delegates to the Third National Congress of the New-Democratic Youth League of China (May 25, 1957).*

A well-disciplined Party armed with the theory of Marxism-Leninism, using the method of self-criticism and linked with the masses of the people; an army under the leadership of such a Party; a united front of all revolutionary classes and all revolutionary groups under the leadership of such a Party—these are the three main weapons with which we have defeated the enemy.
   —*"On the People's Democratic Dictatorship" (June 30, 1949), Selected Works, Vol. IV, p. 422.*

We must have faith in the masses and we must have faith in the Party. These are two cardinal principles. If we doubt these principles, we shall accomplish nothing.
   —*On the Question of Agricultural Co-operation (July 31, 1955), 3rd ed., p. 7.*

---

[4]*running dogs:* Chinese or others who serve the enemy in the way dogs follow a hunter and do his bidding. The Chinese term, *zougou,* is sometimes translated as "lackeys."

2. Classes and Class Struggle

Classes struggle, some classes triumph, others are eliminated. Such is history, such is the history of civilization for thousands of years. To interpret history from this viewpoint is historical materialism; standing in opposition to this viewpoint is historical idealism.
—*"Cast Away Illusions, Prepare for Struggle" (August 14, 1949)*, Selected Works, *Vol. IV, p. 428.*

In class society everyone lives as a member of a particular class, and every kind of thinking, without exception, is stamped with the brand of a class.
—*"On Practice" (July 1937)*, Selected Works, *Vol. I, p. 296.*

Changes in society are due chiefly to the development of the internal contradictions in society, that is, the contradiction between the productive forces and the relations of production, the contradiction between classes and the contradiction between the old and the new; it is the development of these contradictions that pushes society forward and gives the impetus for the supersession of the old society by the new.
—*"On Contradiction" (August 1937)*, Selected Works, *Vol. I, p. 314.*

The ruthless economic exploitation and political oppression of the peasants by the landlord class forced them into numerous uprisings against its rule. . . . It was the class struggles of the peasants, the peasant uprisings and peasant wars that constituted the real motive force of historical development in Chinese feudal society.
—*"The Chinese Revolution and the Chinese Communist Party" (December 1939)*, Selected Works, *Vol. II, p. 308.*

In the final analysis, national struggle is a matter of class struggle. Among the whites in the United States it is only the reactionary ruling circles who oppress the black people. They can in no way represent the workers, farmers, revolutionary intellectuals and other enlightened persons who comprise the overwhelming majority of the white people.
—*"Statement Supporting the American Negroes in Their Just Struggle Against Racial Discrimination by U.S. Imperialism" (August 8, 1963)*, People of the World, Unite and Defeat the U.S. Aggressors and All Their Lackeys, *2nd ed., pp. 3–4.*

31. Women

A man in China is usually subjected to the domination of three systems of authority [political authority, clan authority and religious authority].... As for women, in addition to being dominated by these three systems of authority, they are also dominated by the men (the authority of the husband). These four authorities—political, clan, religious and masculine—are the embodiment of the whole feudal-patriarchal ideology and system, and are the four thick ropes binding the Chinese people, particularly the peasants. How the peasants have overthrown the political authority of the landlords in the countryside has been described above. The political authority of the landlords is the backbone of all the other systems of authority. With that overturned, the clan authority, the religious authority and the authority of the husband all begin to totter.... As to the authority of the husband, this has always been weaker among the poor peasants because, out of economic necessity, their womenfolk have to do more manual labour than the women of the richer classes and therefore have more say and greater power of decision in family matters. With the increasing bankruptcy of the rural economy in recent years, the basis for men's domination over women has already been undermined. With the rise of the peasant movement, the women in many places have now begun to organize rural women's associations; the opportunity has come for them to lift up their heads, and the authority of the husband is getting shakier every day. In a word, the whole feudal-patriarchal ideology and system is tottering with the growth of the peasants' power.

—*"Report on an Investigation of the Peasant Movement in Hunan"* *(March 1927)*, Selected Works, *Vol. I, pp. 44–46.*

Unite and take part in production and political activity to improve the economic and political status of women.

—*Inscription for the magazine* Women of New China, *printed in its first issue, July 20, 1949.*

Protect the interests of the youth, women and children—provide assistance to young students who cannot afford to continue their studies, help the youth and women to organize in order to participate on an equal footing in all work useful to the war effort and to social progress, ensure freedom of marriage and equality as between

men and women, and give young people and children a useful education. . . .
—*"On Coalition Government" (April 24, 1945)*, Selected Works, *Vol. III, p. 288.*

[In agricultural production] our fundamental task is to adjust the use of labour power in an organized way and to encourage women to do farm work.
—*"Our Economic Policy" (January 23, 1934)*, Selected Works, *Vol. I, p. 142.*

In order to build a great socialist society, it is of the utmost importance to arouse the broad masses of women to join in productive activity. Men and women must receive equal pay for equal work in production. Genuine equality between the sexes can only be realized in the process of the socialist transformation of society as a whole.
—*Introductory note to "Women Have Gone to the Labour Front" (1955)*, The Socialist Upsurge in China's Countryside, *Chinese ed., Vol. I.*

With the completion of agricultural co-operation, many co-operatives are finding themselves short of labour. It has become necessary to arouse the great mass of women who did not work in the fields before to take their place on the labour front. . . . China's women are a vast reserve of labour power. This reserve should be tapped in the struggle to build a great socialist country.
—*Introductory note to "Solving the Labour Shortage by Arousing the Women to Join in Production" (1955)*, The Socialist Upsurge in China's Countryside, *Chinese ed., Vol. II.*

Enable every women who can work to take her place on the labour front, under the principle of equal pay for equal work. This should be done as quickly as possible.
—*Introductory note to "On Widening the Scope of Women's Work in the Agricultural Co-operative Movement" (1955)*, The Socialist Upsurge in China's Countryside, *Chinese ed., Vol. I.*

33. Study

In order to have a real grasp of Marxism, one must learn it not only from books, but mainly through class struggle, through practical work and close contact with the masses of workers and peasants. When in addition to reading some Marxist books our intellectuals have gained some understanding through close contact with the masses of workers and peasants and through their own practical work, we will all be speaking the same language, not only the common language of patriotism and the common language of the socialist system, but probably even the common language of the communist world outlook. If that happens, all of us will certainly work much better.

—Speech at the Chinese Communist Party's National Conference on Propaganda Work *(March 12, 1957), 1st pocket ed., p. 12.*

JUST A FEW WORDS, OCTOBER 25, 1966

I have just a few words to say about two matters.

For the past seventeen years there is one thing which in my opinion we haven't done well. Out of concern for state security and in view of the lessons of Stalin in the Soviet Union, we set up a first and second line.[5] I have been in the second line, other comrades in the first line. Now we can see that wasn't so good; as a result our forces were dispersed. When we entered the cities we could not centralize our efforts, and there were quite a few independent kingdoms. Hence the Eleventh Plenum carried out changes. This is one matter. I am in the second line, I do not take charge of day-to-day work. Many things are left to other people so that other people's prestige is built up, and when I go to see God there won't be such a big upheaval in the State. Everybody was in agreement with this idea of mine. It seems that there are some things which the comrades in the first line have not managed too well. There are some things I should have kept a grip on which I did not. So I am responsible, we cannot just blame them. Why do I say that I bear some responsibility?

First, it was I who proposed that the Standing Committee be divided into two lines and that a secretariat be set up. Everyone agreed with this. Moreover I put too much trust in others. It was at the time of the Twenty-three Articles that my vigilance was aroused.* I could do nothing in Beijing; I could do nothing at the Center. Last

---

[5]*First line* means active leadership of the country, and *second line* means serving as top leader for general policy issues but not day-to-day administration. This was set up in China by Mao and Liu Shaoqi to provide an orderly transition of power and to avoid the tribulations of Stalin's last years in the Soviet Union.

*When Mao put forward his new twenty-three-point directive for the Socialist Education Campaign in January 1965 and Liu Shaoqi refused to accept it.

September and October I asked, If revisionism appeared at the Center, what could the localities do? I felt that my ideas couldn't be carried out in Beijing. Why was the criticism of Wu Han[6] initiated not in Beijing but in Shanghai? Because there was nobody to do it in Beijing. Now the problem of Beijing has been solved.

Second, the Great Cultural Revolution wreaked havoc after I approved Nie Yuanzi's big-character poster in Beijing University and wrote a letter to Qinghua University Middle School, as well as writing a big-character poster of my own.* It all happened within a very short period, less than five months in June, July, August, September, and October. No wonder the comrades did not understand too much. The time was so short and the events so violent. I myself had not foreseen that as soon as the Beijing University poster was broadcast, the whole country would be thrown into turmoil. Even before the letter to the Red Guards had gone out, Red Guards had mobilized throughout the country, and in one rush they swept you off your feet. Since it was I who caused the havoc, it is understandable if you have some bitter words for me. Last time we met I lacked confidence and I said that our decisions would not necessarily be carried out. Indeed all that time quite a few comrades still did not understand things fully, though now after a couple of months we have had some experience, and things are a bit better. This meeting has had two stages. In the first stage the speeches were not quite normal, but during the second stage, after speeches and the exchange of experience by comrades at the Center, things went more smoothly and the ideas were understood a bit better. It has only been five months. Perhaps the movement may last another five months, or even longer....

There have been quite a few brief reports presented at this meeting. I have read nearly all of them. You find it difficult to cross this pass and I don't find it easy either. You are anxious and so am I. I can-

---

[6]The criticism of Wu Han, a Beijing historian and official, in late 1965 started the series of events that opened the Cultural Revolution. Wu Han was criticized by radicals for opposing Mao, and the mayor of Beijing, Peng Zhen, was then denounced as Wu Han's protector.

*A translation of the big-character poster "What Are Song Shuo, Lu Ping, and Peng Peiyun Up To in the Great Cultural Revolution?" is in *Survey of China Mainland Press*, no. 3719, 16 June 1966, 6–8. A translation of Mao's "Letter to the Red Guards of Qinghua University Middle School" is in Stuart R. Schram, ed., *Mao Tse-tung Unrehearsed: Talks and Letters 1956–1971* (Harmondsworth: Penguin, 1974), 260–61. A translation of Mao's big-character poster "Bombard the Headquarters" appears earlier in this document.

not blame you, comrades, time has been so short. Some comrades say that they did not intentionally make mistakes, but did it because they were confused. This is pardonable. Nor can we put all the blame on Comrade Shaoqi and Comrade Xiaoping.[7] They have some responsibility, but so has the Center. The Center has not run things properly. The time was so short. We were not mentally prepared for new problems. Political and ideological work was not carried out properly. I think that after this seventeen-day conference things will be a bit better.

Does anyone else want to speak? I guess that's all for today. The meeting is adjourned.

---

[7] *Comrade Xiaoping:* Deng Xiaoping, who became China's paramount leader after Mao (1977–1997)

PART TWO

Documenting Mao

11

EDGAR SNOW
Interview with Mao
1937

Edgar Snow's interviews with Mao in 1936 have the status of a classic in both China and the West. These were the first public comments by Mao about himself and his life. Snow, a progressive American journalist, had lived and worked in China since 1929. In June 1936, he set out to visit the "Reds" who had recently made their famous Long March from southeast China to the northwest. Snow published his observations in a book, Red Star over China, *that became an instant bestseller in the West and was quickly translated into Chinese. Although the book includes his impressions of the Communist administration in general, it is his interviews with Mao that have always attracted the most attention.*

Mao proves himself a brilliant storyteller, and Snow a faithful transmitter. The selections from Red Star over China *printed here include Snow's memorable impressions of Mao as a man and portions of Mao's own story about his youth and the activities of the Red Army in the late 1920s and early 1930s. In all, Mao manages to weave his story with that of the CCP, thus beginning his ultimately successful effort to make himself the embodiment of the party.*

. . . Here at last I found the Red leader whom Nanjing had been fighting for ten years—Mao Zedong, chairman of the "Chinese People's Soviet Republic," to employ the official title which had recently been adopted. The old cognomen, "Chinese Workers' and Peasants' Soviet Republic," was dropped when the Reds[1] began their new policy of struggle for a united front. . . .

Mao had the reputation of a charmed life. He had been repeatedly pronounced dead by his enemies, only to return to the news columns a few days later, as active as ever. The Guomindang had also officially

[1]*Reds:* the common name for the Communists at that time

Edgar Snow, *Red Star over China* (New York: Grove Press, 1968), 89–96, 130–33, 172–74.

"killed" and buried Zhu De[2] many times, assisted by occasional corroborations from clairvoyant missionaries. Numerous deaths of the two famous men, nevertheless, did not prevent them from being involved in many spectacular exploits, including the Long March. Mao was indeed in one of his periods of newspaper demise when I visited Red China,[3] but I found him quite substantially alive. There were good reasons why people said that he had a charmed life, however; although he had been in scores of battles, was once captured by enemy troops and escaped, and had the world's highest reward on his head, during all these years he had never once been wounded.

I happened to be in Mao's house one evening when he was given a complete physical examination by a Red surgeon—a man who had studied in Europe and who knew his business—and pronounced in excellent health. He had never had tuberculosis or any "incurable disease," as had been rumored by some romantic travelers. His lungs were completely sound, although, unlike most Red commanders, he was an inordinate cigarette smoker. During the Long March, Mao and Li De[4] had carried on original botanical research by testing out various kinds of leaves as tobacco substitutes. . . .

Mao Zedong was forty-three years old when I met him in 1936. He was elected chairman of the provisional Central Soviet Government at the Second All-China Soviet Congress, attended by delegates representing approximately 9,000,000 people then living under Red laws. . . .

The influence of Mao Zedong throughout the Communist world of China was probably greater than that of anyone else. He was a member of nearly everything—the revolutionary military committee, the political bureau of the Central Committee, the finance commission, the organization committee, the public health commission, and others. His real influence was asserted through his domination of the political bureau, which had decisive power in the policies of the Party, the government, and the army. Yet, while everyone knew and respected him, there was—as yet, at least—no ritual of hero worship built up around him. I never met a Chinese Red who drooled "our-great-leader" phrases; I did not hear Mao's name used as a synonym for the

[2]*Zhu De:* founder of the Red Army who fought alongside Mao in southern China and became one of the most respected leaders of the new CCP government
[3]*Red China:* the base area in northwest China that was controlled by the CCP and would soon be known as Yan'an
[4]*Li De:* Otto Braun, an adviser from the Comintern, or Communist International, to the CCP from 1933 to 1937; the only foreign participant to complete the Long March

Chinese people, but still I never met one who did not like "the Chairman"—as everyone called him—and admire him. The role of his personality in the movement was clearly immense.

Mao seemed to me a very interesting and complex man. He had the simplicity and naturalness of the Chinese peasant, with a lively sense of humor and a love of rustic laughter. His laughter was even active on the subject of himself and the shortcomings of the soviets— a boyish sort of laughter which never in the least shook his inner faith in his purpose. He was plain-speaking and plain-living, and some people might have considered him rather coarse and vulgar. Yet he combined curious qualities of naïveté with incisive wit and worldly sophistication.

I think my first impression—dominantly one of native shrewdness—was probably correct. And yet Mao was an accomplished scholar of Classical Chinese, an omnivorous reader, a deep student of philosophy and history, a good speaker, a man with an unusual memory and extraordinary powers of concentration, an able writer, careless in his personal habits and appearance but astonishingly meticulous about details of duty, a man of tireless energy, and a military and political strategist of considerable genius. It was interesting that many Japanese regarded him as the ablest Chinese strategist alive.

The Reds were putting up some new buildings in Bao'an,[5] but accommodations were very primitive while I was there. Mao lived with his wife in a two-room yaofang[6] with bare, poor, map-covered walls. He had known much worse, and as the son of a "rich" peasant in Hunan he had also known better. The Maos' chief luxury was a mosquito net. Otherwise Mao lived very much like the rank and file of the Red Army. After ten years of leadership of the Reds, after hundreds of confiscations of property of landlords, officials, and tax collectors, he owned only his blankets and a few personal belongings, including two cotton uniforms. Although he was a Red Army commander as well as chairman, he wore on his coat collar only the two red bars that are the insignia of the ordinary Red soldier.

I went with Mao several times to mass meetings of the villagers and the Red cadets, and to the Red theater. He sat inconspicuously in the midst of the crowd and enjoyed himself hugely. I remember once, between acts at the Anti-Japanese Theater, there was a general

[5] *Bao'an:* the town in Shaanxi province where Snow interviewed Mao. In a few months, Mao and the CCP leadership would move to the larger town of Yan'an.
[6] yaofang: a style of cave house common in northern Shaanxi province

demand for a duet by Mao Zedong and Lin Biao, the twenty-eight-year-old president of the Red Army University and formerly a famed young cadet on Chiang Kai-shek's staff. Lin blushed like a schoolboy and got them out of the "command performance" by a graceful speech, calling upon the women Communists for a song instead.

Mao's food was the same as everybody's, but being a Hunanese he had the southerner's *ai-la*, or "love of pepper." He even had pepper cooked into his bread. Except for this passion, he scarcely seemed to notice what he ate. One night at dinner I heard him expand on a theory of pepper-loving peoples being revolutionaries. He first submitted his own province, Hunan, famous for the revolutionaries it has produced. Then he listed Spain, Mexico, Russia, and France to support his contention, but laughingly had to admit defeat when somebody mentioned the well-known Italian love of red pepper and garlic, in refutation of his theory. One of the most amusing songs of the "bandits," incidentally, was a ditty called "The Hot Red Pepper." It told of the disgust of the pepper with his pointless vegetable existence, waiting to be eaten, and how he ridiculed the contentment of the cabbages, spinach, and beans with their invertebrate careers. He ends up by leading a vegetable insurrection. "The Hot Red Pepper" was a great favorite with Chairman Mao.

He appeared to be quite free from symptoms of megalomania, but he had a deep sense of personal dignity, and something about him suggested a power of ruthless decision when he deemed it necessary. I never saw him angry, but I heard from others that on occasions he had been roused to an intense and withering fury. At such times his command of irony and invective was said to be classic and lethal.

I found him surprisingly well informed on current world politics. Even on the Long March, it seems, the Reds received news broadcasts by radio, and in the Northwest they published their own newspapers. Mao was exceptionally well read in world history and had a realistic conception of European social and political conditions. He was very interested in the Labour Party of England, and questioned me intensely about its present policies, soon exhausting all my information. It seemed to me that he found it difficult fully to understand why, in a country where workers were enfranchised, there was still no workers' government. I was afraid my answers did not satisfy him. He expressed profound contempt for Ramsay MacDonald,[7] whom he designated as a *han-chien* — an archtraitor of the British people.

[7]*Ramsay MacDonald:* former British prime minister. MacDonald had founded the Labour party and promoted the sort of parliamentary and moderate socialism that Mao felt betrayed the revolutionary interests of the working class.

His opinion of President Roosevelt was rather interesting. He believed him to be anti-Fascist, and thought China could cooperate with such a man. He asked innumerable questions about the New Deal,[8] and Roosevelt's foreign policy. The questioning showed a remarkably clear conception of the objectives of both. He regarded Mussolini and Hitler as mountebanks, but considered Mussolini intellectually a much abler man, a real Machiavellian, with a knowledge of history, while Hitler was a mere willless puppet of the reactionary capitalists.[9]

Mao had read a number of books about India and had some definite opinions on that country. Chief among these was that Indian independence would never be realized without an agrarian revolution. He questioned me about Gandhi, Jawaharlal Nehru, Suhasini Chattopadhyaya,[10] and other Indian leaders I had known. He knew something about the Negro question in America, and unfavorably compared the treatment of Negroes and American Indians with policies in the Soviet Union toward national minorities. He was interested when I pointed out certain great differences in the historical background of the Negro in America and that of minorities in Russia.

Mao was an ardent student of philosophy. Once when I was having nightly interviews with him on Communist history, a visitor brought him several new books on philosophy, and Mao asked me to postpone our engagements. He consumed those books in three or four nights of intensive reading, during which he seemed oblivious to everything else. He had not confined his reading to Marxist philosophers, but also knew something of the ancient Greeks, of Spinoza, Kant, Goethe, Hegel, Rousseau, and others.

I often wondered about Mao's own sense of responsibility over the question of force, violence, and the "necessity of killing." He had in his youth had strongly liberal and humanistic tendencies, and the transition from idealism to realism evidently had first been made philosophically. Although he was peasant-born, he did not as a youth personally suffer much from oppression of the landlords, as did many Reds, and, although Marxism was the core of his thought, I deduced that class hatred was for him probably an intellectually acquired mechanism in the bulwark of his philosophy, rather than an instinctive impulse to action.

[8]*New Deal:* the set of economic and social policies pursued by Roosevelt in the 1930s to combat the Depression

[9]Hitler and Mussolini were by 1936 the fascist leaders of Germany and Italy. Mao was not alone in thinking that Mussolini was the more significant of the two.

[10]Jawaharlal Nehru and Suhasini Chattopadhyaya were prominent leaders of the nationalist movement in India, which was seeking independence from British rule. Nehru became independent India's first prime minister in 1947.

There seemed to be nothing in him that might be called religious feeling. He was a humanist in a fundamental sense; he believed in man's ability to solve man's problems. I thought he had probably on the whole been a moderating influence in the Communist movement where life and death were concerned. . . .

Yet I doubted very much if he would ever command great respect from the intellectual elite of China, perhaps not entirely because he had an extraordinary mind, but because he had the personal habits of a peasant. The Chinese disciples of Pareto[11] might have thought him uncouth. Talking with Mao one day, I saw him absent-mindedly turn down the belt of his trousers and search for some guests[12]—but then it is just possible that Pareto might have done a little searching himself if he had lived in similar circumstances. But I am sure that Pareto would never have taken off his trousers in the presence of the president of the Red Army University—as Mao did once when I was interviewing Lin Biao. It was extremely hot inside the little cave. Mao lay down on the bed, pulled off his pants, and for twenty minutes carefully studied a military map on the wall—interrupted occasionally by Lin Biao, who asked for confirmation of dates and names, which Mao invariably knew. His nonchalant habits fitted with his complete indifference to personal appearance, although the means were at hand to fix himself up like a chocolate-box general or a politician's picture in *Who's Who in China.*

Except for a few weeks when he was ill, he walked most of the 6,000 miles of the Long March, like the rank and file. He could have achieved high office and riches by "betraying" to the Guomindang, and this applied to most Red commanders. The tenacity with which these Communists for ten years clung to their principles could not be fully evaluated unless one knew the history of "silver bullets"[13] in China, by means of which other rebels were bought off.

I was able to check up on many of Mao's assertions, and usually found them to be correct. He subjected me to mild doses of political propaganda, but it was interesting compared to what I had received in nonbandit quarters. He never imposed any censorship on me, in either my writing or my photography, courtesies for which I was grateful. He did his best to see that I got facts to explain various aspects of soviet life.

[11]*Pareto:* Vilfredo Pareto, the noted leftist theorist who critiqued parliamentary democracy and promoted Marx's vision of class struggle but was a thoroughly bourgeois professor in Switzerland
[12]*search for some guests:* hunt for lice
[13]*silver bullets:* bribes

[Mao gave his story to Edgar Snow, who recorded it as follows.]

"I was born in the village of Shaoshan, in Xiangtan county, Hunan province, in 1893. My father's name was Mao Rensheng, and my mother's maiden name was Wen Qimei.

"My father was a poor peasant and while still young was obliged to join the army because of heavy debts. He was a soldier for many years. Later on he returned to the village where I was born, and by saving carefully and gathering together a little money through small trading and other enterprise he managed to buy back his land. . . .

"I began studying in a local primary school when I was eight and remained there until I was thirteen years old. In the early morning and at night I worked on the farm. During the day I read the Confucian Analects and the Four Classics.[14] My Chinese teacher belonged to the stern-treatment school. He was harsh and severe, frequently beating his students. Because of that I ran away from the school when I was ten. I was afraid to return home for fear of receiving a beating there, and set out in the general direction of the city, which I believed to be in a valley somewhere. I wandered for three days before I was finally found by my family. Then I learned that I had circled round and round in my travels, and in all my walking had got only about eight *li* from my home.

"After my return to the family, however, to my surprise conditions somewhat improved. My father was slightly more considerate and the teacher was more inclined to moderation. The result of my act of protest impressed me very much. It was a successful 'strike.'

"My father wanted me to begin keeping the family books as soon as I had learned a few characters. He wanted me to learn to use the abacus. As my father insisted upon this I began to work at those accounts at night. He was a severe taskmaster. He hated to see me idle, and if there were no books to be kept he put me to work at farm tasks. He was a hot-tempered man and frequently beat both me and my brothers. He gave us no money whatever, and the most meager food. On the fifteenth of every month he made a concession to his laborers and gave them eggs with their rice, but never meat. To me he gave neither eggs nor meat.

"My mother was a kind woman, generous and sympathetic, and ever ready to share what she had. She pitied the poor and often gave them rice when they came to ask for it during famines. But she could

---

[14]The *Analects* is the record of the teachings of Confucius. Since the twelfth century, it has been considered one of the Four Books that all scholars should study, the other three being *The Mencius, The Great Learning,* and *The Doctrine of the Mean.*

not do so when my father was present. He disapproved of charity. We had many quarrels in my home over this question.

"There were two 'parties' in the family. One was my father, the Ruling Power. The Opposition was made up of myself, my mother, my brother, and sometimes even the laborer. In the 'united front' of the Opposition, however, there was a difference of opinion. My mother advocated a policy of indirect attack. She criticized any overt display of emotion and attempts at open rebellion against the Ruling Power. She said it was not the Chinese way.

"But when I was thirteen I discovered a powerful argument of my own for debating with my father on his own ground, by quoting the Classics. My father's favorite accusations against me were of unfilial conduct and laziness. I quoted, in exchange, passages from the Classics saying that the elder must be kind and affectionate. Against his charge that I was lazy I used the rebuttal that older people should do more work than younger, that my father was over three times as old as myself, and therefore should do more work. And I declared that when I was his age I would be much more energetic.

"The old man continued to 'amass wealth,' or what was considered to be a great fortune in that little village. He did not buy more land himself, but he bought many mortgages on other people's land. His capital grew to two or three thousand Chinese dollars.*

"My dissatisfaction increased. The dialectical struggle in our family was constantly developing.† One incident I especially remember. When I was about thirteen my father invited many guests to his home, and while they were present a dispute arose between the two of us. My father denounced me before the whole group, calling me lazy and useless. This infuriated me. I cursed him and left the house. My mother ran after me and tried to persuade me to return. My father also pursued me, cursing at the same time that he commanded me to come back. I reached the edge of a pond and threatened to jump in if he came any nearer. In this situation demands and counterdemands were presented for cessation of the civil war. My father insisted that I apologize and *koutou*‡ as a sign of submission. I agreed to give a one-knee *koutou* if he would promise not to beat me. Thus the war ended,

---

*Mao used the Chinese term *yuan*, which was often translated as "Chinese dollars"; 3,000 yuan in cash in 1900 was an impressive sum in rural China.
†Mao used all these political terms humorously in his explanations, laughing as he recalled such incidents.
‡Literally, to "knock head." To strike one's head to the floor or earth was expected of son to father and subject to emperor, in token of filial obedience.

and from it I learned that when I defended my rights by open rebellion my father relented, but when I remained meek and submissive he only cursed and beat me the more.

"Reflecting on this, I think that in the end the strictness of my father defeated him. I learned to hate him, and we created a real united front against him. At the same time it probably benefited me. It made me most diligent in my work; it made me keep my books carefully, so that he should have no basis for criticizing me. . . ."

[Snow resumes Mao's story.]

"Gradually the Red Army's work with the masses improved, discipline strengthened, and a new technique in organization developed. The peasantry everywhere began to volunteer to help the revolution. As early as Jingganshan[15] the Red Army had imposed three simple rules of discipline upon its fighters, and these were: prompt obedience to orders; no confiscations whatever from the poor peasantry; and prompt delivery directly to the government, for its disposal, of all goods confiscated from the landlords. After the 1928 Conference [second Maoping Conference] emphatic efforts to enlist the support of the peasantry were made, and eight rules were added to the three listed above. These were as follows:

"1. Replace all doors when you leave a house;*
"2. Return and roll up the straw matting on which you sleep;
"3. Be courteous and polite to the people and help them when you can;
"4. Return all borrowed articles;
"5. Replace all damaged articles;
"6. Be honest in all transactions with the peasants;
"7. Pay for all articles purchased;
"8. Be sanitary, and, especially, establish latrines a safe distance from people's houses.

"The last two rules were added by Lin Biao. These eight points were enforced with better and better success, and today are still

---

[15]*Jingganshan:* the CCP's first substantial base area, formed in the rugged mountains of rural Jiangxi province in southern China. Mao and Zhu De retreated there in late 1927 after the failure of urban insurrections and together developed the Red Army and basic approaches to running a rural soviet.

*This order is not so enigmatic as it sounds. The wooden doors of a Chinese house are easily detachable, and are often taken down at night, put across wooden blocks, and used for an improvised bed.

the code of the Red soldier, memorized and frequently repeated by him. Three other duties were taught to the Red Army, as its primary purpose: first, to struggle to the death against the enemy; second, to arm the masses; third, to raise money to support the struggle. . . .

"It was at this time that the First Army Corps was organized, with Zhu De as commander and myself as political commissar. It was composed of the Third Army, the Fourth Army commanded by Lin Biao, and the Twelfth Army, under Luo Pinghui. Party leadership was vested in a Front Committee, of which I was chairman. There were already more than 10,000 men in the First Army Corps then, organized into ten divisions. Besides this main force, there were many local and independent regiments, Red Guards and partisans.

"Red tactics, apart from the political basis of the movement, explained much of the successful military development. At Jingganshan four slogans had been adopted, and these give the clue to the methods of partisan warfare used, out of which the Red Army grew. The slogans were:

"1. When the enemy advances, we retreat!
"2. When the enemy halts and encamps, we trouble them!
"3. When the enemy seeks to avoid a battle, we attack!
"4. When the enemy retreats, we pursue!

"These slogans [of four characters each in Chinese] were at first opposed by many experienced military men, who did not agree with the type of tactics advocated. But much experience proved that the tactics were correct. Whenever the Red Army departed from them, in general, it did not succeed. Our forces were small, exceeded from ten to twenty times by the enemy; our resources and fighting materials were limited, and only by skillfully combining the tactics of maneuvering and guerrilla warfare could we hope to succeed in our struggle against the Guomindang, fighting from vastly richer and superior bases.

"The most important single tactic of the Red Army was, and remains, its ability to concentrate its main forces in the attack, and swiftly divide and separate them afterwards. This implied that positional warfare was to be avoided, and every effort made to meet the living forces of the enemy while in movement, and destroy them. On the basis of these tactics the mobility and the swift, powerful 'short attack' of the Red Army was developed.

12

STUART SCHRAM

The Struggle on Two Fronts
1967

Stuart Schram is the dean of Mao studies in the West. Beginning in the mid-1960s, he produced an excellent biography of Mao, from which this brief selection is taken, as well as several editions of selected translations of Mao's writings and other studies on Mao and modern China. Although many biographies of Mao have been written in the past thirty-five years, Schram's has retained its reputation among scholars as one of the most reliable and well researched.

This selection describes the context out of which the famous Yan'an Rectification Movement emerged. Mao and the CCP faced a struggle on two fronts—against the Japanese invaders and against their erstwhile allies, the Nationalists, under Chiang Kai-shek. Both wanted to see the end of the CCP. Mao's response was ideological and educational, as well as military and organizational. This was the context of Mao's successful effort to Sinify (adapt to Chinese conditions) Russian Marxism-Leninism. (See Nick Knight's analysis of Mao's "Sinification of Marxism" in Document 13.) No matter how one regrets the excesses of Mao's later career, Schram's biography gives a vivid sense of what Mao and the CCP had to overcome to reunify the nation.

The years in the North-West saw Mao's emergence as a public figure, both in China and abroad. Though often mentioned in the press during the previous decade, he had remained little more than a name: a bandit chieftain for some, a revolutionary hero for others, but in either case an abstraction. Now, following Edgar Snow's pioneering journey in the summer of 1936, a succession of visitors contrived to reach the "Border Region," and to bring back first-hand reports on its inhabitants.

The portraits they painted of Mao were varied, except in one respect: he left no one indifferent. Edgar Snow, the first foreign

Stuart R. Schram, *Mao Tse-tung* (Harmondsworth: Penguin, 1967), 209–20. I have edited Schram's footnotes lightly to limit scholarly detail.

journalist to interview him, described Mao justly as characterized by a combination of intellectual depth and peasant shrewdness. Though the cult of the leader was not to develop seriously for another five years, Snow noted Mao's "deep sense of personal dignity," as well as his "power of ruthless decision." One of the most perceptive accounts of Mao's personality has been left by an intense and highly sensitive woman, Agnes Smedley. She found him, at their first meeting, effeminate and vaguely repellent. "Whatever else he might be," she wrote, "he was an aesthete." Later she discovered the strength behind the mobile features and graceful gestures, and the human warmth and simplicity behind the dignified reserve. But though he could communicate intensely with a few intimate friends, he remained on the whole reserved and aloof. "The sinister quality I had at first felt so strongly in him," continued Agnes Smedley, "proved to be a spiritual isolation. As Zhu De was loved, Mao Zedong was respected. The few who came to know him best had affection for him, but his spirit dwelt within itself, isolating him.". . .

The Yan'an period, especially the two years from the beginning of 1938 to the beginning of 1940, is an altogether exceptional one in Mao's literary career. Not only was his total output very high, but it consisted more of relatively long and systematic writings and less of reports, speeches, and directives than at any other time in his life. It is not difficult to understand the reasons for this. On the one hand, the stable situation in the base area resulting from the accord with the Guomindang, and the fact that he was not personally involved in combat against the Japanese, gave him more leisure than he had enjoyed since his student days, except perhaps for a year or two during the heyday of the Jiangxi Soviet Republic. And on the other hand, he was at last achieving the grasp of Marxist theory, the self-confidence, and the breadth of vision necessary to deal globally with the problems of the Chinese revolution. . . .

In his analysis of the anti-Japanese war as a whole Mao envisaged three phases. During the first, which was nearly completed, the Japanese would be on the offensive and the Chinese on the defensive. During the second, which would be the longest, there would be a certain equilibrium, and the Chinese would carry on guerrilla warfare behind the Japanese lines. During the third and last phase the guerrilla units which had been organized and forged in the course of the preceding phases would swing over to the offensive and abandon their guerrilla tactics for conventional mobile warfare on a large scale. But final victory would be possible only if their efforts were supported both by the conventional forces of the Nationalist army, and by other democratic powers opposed to Japanese imperialism.

In the uncertain and rapidly changing international situation on the eve of the Second World War it was difficult for Mao, or for anyone else, to know exactly on whose support China could count. As late as the beginning of 1939 he professed to believe that, despite Chamberlain's policy of appeasement[1] and his "cowardly attitude towards Japan," the "broad popular masses, including all progressive people from the various social strata who sympathize with China," would in the end succeed in convincing the governments of England, France, and America that it was in their own interests to oppose Japanese aggression. In September 1939, following the Nazi-Soviet pact,[2] he condemned all the "imperialist powers" with equal violence as "mad dogs" who could not do otherwise than "hurl themselves pell mell against their enemies and against the walls of the world." He even suggested that Chamberlain, the leader of "the most reactionary country in the world," was worse than Hitler....

To be sure, the development of the guerrilla base areas was the most effective manner in which Mao and his comrades could fight the Japanese. But Chiang Kai-shek could not be blind to the enormous increase in the Communists' power which resulted from these activities, and which would fortify their bargaining position with the Nationalists once the invaders were expelled. As Chalmers Johnson has put it: "Although the Communist Army was actually continuing to fight the Japanese, its methods left no room for more than one Chinese victor over Japan."*

... This judgement in fact sums up the whole subsequent course of the anti-Japanese war, during which Mao and his comrades ceased to make even a pretence of obeying Guomindang directives, yet continuously increased their prestige as the staunchest defenders of China's national interests.

At the same time that their relations with the Guomindang thus deteriorated the Communists found themselves under increasingly fierce attack by the Japanese. The latter, who had already been building blockhouses to separate and isolate the guerrilla areas, intensified

---

[1]Schram is referring to British prime minister Neville Chamberlain, who turned a blind eye to Japanese aggression in China.

[2]*Nazi-Soviet pact:* the nonaggression pact between Germany and the Soviet Union, signed August 23, 1939.

*Chalmers Johnson, *Peasant Nationalism and Communist Power* (Stanford: Stanford University Press, 1962).

this policy in order to prevent any recurrence of the Hundred Regiments' Offensive.[3] In July 1941 the notorious "three-all policy" (burn all, kill all, loot all) was adopted in North China. The result was to reduce the size of the Eighth Route Army from 400,000 to 300,000 men, and to diminish the population of the Communist areas in North China from 44 millions to 25 millions.

The mass burning and killing associated with the "three-all" policy on the whole intensified the peasants' will to resist, and thus aided the Communist cause. But at the same time the Japanese construction of blockhouses, walls, and barbed-wire enclosures, with the virtually total economic blockade of the Communist areas henceforth imposed by the Guomindang, produced increasingly difficult economic and political problems for Mao and his comrades. The economic challenge was met by a successful drive to increase production in the base areas themselves, involving the encouragement of mutual-aid teams among the peasantry, and the participation of the army in industrial and agricultural work. The political challenge was met by the "Zhengfeng" or rectification campaign, which was formally launched in February 1942.

The rectification campaign had a dual purpose, neither aspect of which should be ignored. On the one hand, it was designed to strengthen the unity and discipline of the Chinese Communist Party in difficult circumstances, imparting a minimum knowledge of Marxism-Leninism and the political methods developed in the Soviet Union to the large number of new recruits who had been taken into the party in the course of its wartime expansion. But it was also designed to give to the ideological consciousness of party members a special and characteristic quality directly inspired by Mao Zedong himself. In short, Mao's aim was "sinification" of Marxism. . . .

Launching the rectification campaign in February 1942, Mao once more energetically denounced party formalism in general, and "foreign formalism" in particular. In the second of his two long speeches he complained that his appeal of 1938 for the adaptation of Marxism to Chinese conditions had not been heard, and demanded that it be put into practice immediately.

The previous May Mao had criticized only the students returned from Europe, America, or Japan; but now it was clear that the "foreign formalism" he desired to eliminate consisted in the imitation of Soviet models. The rectification campaign was thus directed in part against

[3] *Hundred Regiments Offensive:* the surprise Communist attack on Japanese forces in northern China in the summer of 1940. At first it was a great success, turning back Japanese forces at the margin, but it brought a cruel series of counteroffensives in the fall of 1940 and all of 1941 that nearly crushed the CCP.

the predominance of Soviet influence in the party. The most recent official history of the Chinese Communist Party makes this quite clear when it explains that the campaign was aimed at "doctrinaires as represented by Comrade Wang Ming" who were "ignorant of the Party's historical experience," and could "only quote words or phrases from Marxist writings."

To be sure, the fact that the rectification campaign was designed to make the Chinese Communists more aware of their own history and traditions, and teach them to deal with the problems of the Chinese revolution in an independent and original manner, does not mean that it was, properly speaking, *anti*-Soviet. Approximately one fourth of the materials eventually put together in a volume for study by party members consisted of translations from Lenin, Stalin, and Dimitrov.* But if the Chinese Communists were to be instructed in the ideas and methods of their Soviet comrades, they were instructed as well to regard all such with detachment, to choose and assimilate only what was useful to them. A fully autonomous Communist movement in China, with its own ideology specifically adapted to Chinese conditions, was Mao's goal.

*See Boyd Compton ed., *Mao's China: Party Reform Documents, 1942–1944* (Seattle: University of Washington Press, 1952).

13

NICK KNIGHT

Mao Zedong's "Sinification of Marxism"

1985

Nick Knight's essay is a philosophical study of Mao's "theory of practice" that helps us see why Mao's version of Marxism made sense to many Chinese as a social theory and a guide to action—not only during World War II but even today. Sinification *means making something from outside China adapt or integrate into Chinese society and thought. Mao did this with Communist ideology—Marxism-Leninism—in the Yan'an*

Nick Knight, "Mao Zedong and the 'Sinification of Marxism'" in *Marxism in Asia,* ed. Colin Mackerras and Nick Knight (New York: St. Martin's Press, 1985), 83–90.

period (1936–47). Knight is a senior Mao scholar among Western China specialists, and his empathetic analysis of Mao's use of "particular laws" and "universal laws" (of nature, history, and how to wage war) reveals the reasonableness of Mao's ideas on how to make a revolution. Knight's chart on page 203 maps out the cycle of theory-practice-theory presented in Mao's "Methods of Leadership" (1943, see Document 4). When the CCP carefully applied this method, it was often quite successful—both in gaining power and in improving the lives of most people in a given locality. The tragedy is that Mao and the CCP soon abandoned this time-consuming method in favor of the simple and disastrous method of dictatorship. Even so, Mao's call to combine theory and practice in the search for social justice remains a compelling ideal that still attracts some Chinese and many revolutionaries elsewhere in the world today.

In . . . October 1938, Mao called for the Sinification of Marxism in the following terms:

> A communist is a Marxist internationalist, but Marxism must take on a national form before it can be applied. There is no such thing as abstract Marxism, but only concrete Marxism. What we call Marxism is Marxism that has taken on a national form, that is, Marxism applied to the concrete struggle in the concrete conditions prevailing in China, and not Marxism abstractly used. If a Chinese communist, who is a part of the great Chinese people, bound to his people by his very flesh and blood, talks of Marxism apart from Chinese peculiarities, this Marxism is merely an empty abstraction. Consequently, the Sinification of Marxism—that is to say, making certain that in all of its manifestations it is imbued with Chinese peculiarities, using it according to these peculiarities—becomes a problem that must be understood and solved by the whole party without delay.*

How is this Sinification of Marxism to be interpreted? Western critiques have tended to cluster around two lines of explication. The first suggests that the Sinification of Marxism entailed the elevation of Chinese tradition and realities at the expense of Marxism's universal truths. Stuart Schram, for example, has argued that Mao denied altogether the existence of "a universally valid form of Marxism," and that

---

*Stuart Schram, *The Political Thought of Mao Tse-tung*, rev. ed. (Harmondsworth: Penguin, 1969), 172.

his "preoccupation with the glory of China" led to a Sinification of Marxism which was "hermetic."* The second argues that the Sinification of Marxism was a ploy used by Mao to enhance his own position in the power struggle with the Moscow-oriented Returned Students' Faction[1] which had favoured a more orthodox European and Soviet reading of Marxism. According to Robert North, Mao was "adapting Russian communist political theory to meet peculiar Chinese requirements and the convenience of his own climb to power."† However, a third explanation is possible: that Mao was attempting to establish a formula by which a universal theory such as Marxism could be utilised in a particular national context and culture *without abandoning the universality of that theory.*‡

How can such an interpretation be justified? It is important to point out that Mao believed it possible to discover particular "laws" (of nature, society, history, war) which unlike the universal laws of Marxism, have no universal applicability. This belief in the existence of particular "laws" describing the regularities present in particular or localised contexts comes across clearly in the following interesting passage from "Problems of Strategy in China's Revolutionary War," written by Mao in 1936:

> ... the different laws for directing wars are determined by the different circumstances of those wars—differences in their time, place and nature *(xingzhi)*. As regards the time factor, both war and its laws develop; each historical stage has its special characteristics, and hence the laws in each historical stage have their special characteristics and cannot be mechanically applied in another stage. As for the nature of war, since revolutionary war and counter-revolutionary war both have their special characteristics, the laws governing them also have their own characteristics, and those applying to one cannot be mechanically transferred to the other. As for the factor of place, since each country or nation, especially a large country or nation, has its own characteristics, the laws of war for each country or nation also have their own characteristics, and

*Ibid., 112–16.

[1] *Returned Students' Faction:* refers to Chinese Communists who studied in Moscow in the 1920s and early 1930s. They became competitors to Mao for leadership of the CCP.

†Robert C. North, *Moscow and Chinese Communists* (Stanford: Stanford University Press, 1953 & 1963), 193.

‡The following analysis is a truncated version of my article "The Form of Mao Zedong's 'Sinification of Marxism,'" *Australian Journal of Chinese Affairs,* no. 9 (January 1983): 17–33.

here, too, those applying to one cannot be mechanically transferred to the other. In studying the laws for directing wars that occur at different historical stages, that differ in nature and that are waged in different places and by different nations, we must fix our attention on the characteristics and development of each, and must oppose a mechanical approach to the problem of war.*

It is evident from this passage that Mao rejected the notion that there can only be laws of war in general. On the contrary, he argues that it is possible and desirable to seek out "laws" describing the characteristics of specific theatres of war.

But what is the relationship between universal laws of history and society and the particular "laws" of Chinese society and the Chinese revolution which Mao was so eager to discover? In the first place, it would seem that universal laws are derived from the many particular "laws" which describe the behaviour of the same phenomenon in different situations. For example, if different forms of class struggle are observed in many different social contexts, the particular "laws" describing those specific instances of class struggle constitute the foundation of a universal law which asserts the existence of class struggle in *all* societies. In the second place, a universal law has a predictive value which asserts the certain existence of the phenomenon described by that law in instances as yet unobserved; the universal law thus draws attention to, and guides analysis of this phenomenon. The function of such a universal law was not, as far as Mao was concerned, to dictate the manner in which that phenomenon would assume concrete form; that could only be disclosed through an empirical investigation whose purpose was to arrive at an understanding of the characteristics of that phenomenon in its specific context. In other words, the particular "law" would be revealed by an empirical investigation informed by knowledge of the relevant universal law. To pursue our example, the universal law of class struggle represented an abstract assertion of the existence of class struggle in all societies (save perhaps the most primitive); knowledge of that universal law had to be mobilised to discover the particular "law" which described the nature of class struggle within Chinese society.

Mao believed that the abstract universal laws of Marxism performed an important function in directing attention to those aspects of society requiring analysis and study. However, such universal laws did *not* by themselves represent Marxism in its entirety. To become a complete ideology, the universal laws of Marxism had to be mobilised

Selected Works of Mao Tse-tung (Peking: Foreign Languages Press, 1967), 1:181–82.

and applied to disclose a society's particular "laws." It was this union of the universal (abstract) laws and the particular (concrete) "laws" which constitute Marxism in its totality. And obviously, such a union of universal and particular could occur only within a specific historical and social context such as the Chinese society during the period of the anti-Japanese war. Consequently, Marxism in the Chinese context consisted of Marxism's universal laws utilised to disclose the particular "laws" describing the nature of Chinese society and the Chinese revolution. Once disclosed, those particular "laws" became for Mao an integral element of Marxism within that historically defined situation. It is in this sense that Mao could call for the Sinification of Marxism.

From this perspective, the Sinification of Marxism was not a question of the elevation of Chinese realities at the expense of Marxism's universality, but the completion of Marxism as an ideological system. Inherent in Mao's Sinification of Marxism is the notion that Marxism as a complete ideological system (rather than just a body of abstract universal laws) is definable only within a concrete historical situation. Thus, although the Sinification of Marxism is, as [Raymond] Wylie claims, a "culturally charged term,"* it does not claim any cultural privilege over Marxism. Within a different cultural or historical context, the universal laws of Marxism would have to be joined with the particular "laws" of that specific situation. Because these particular "laws" would be different from those describing Chinese society, that particular Marxism would differ accordingly. Both, nevertheless, would share a common stock of universal laws.

This view of Marxism led logically to an insistence on the need to pay close attention to the particular characteristics of Chinese society and history. Mao was to return to this point again and again in subsequent writings of the Yan'an period, and he made no attempt to conceal his impatience with those Marxists who were preoccupied with foreign models and history to the exclusion of Chinese history and conditions. He perceived this preoccupation as largely a manifestation of an incorrect interpretation of Marxism, one which regarded the particular "laws" of a largely European form of Marxism as having relevance within the Chinese context. In "Reform Our Study" (1941), Mao isolated three conditions having deleterious effect within the CCP: the study of current conditions was being neglected, as were the study of history and the application of Marxism-Leninism. For Mao,

*Raymond Wylie, *The Emergence of Maoism* (Stanford: Stanford University Press, 1980), 52.

these failings were a manifestation of an incorrect interpretation of Marxism. His critique of them was inspired by his own view of Marxism which insisted on the integration of Marxism's universal laws with the particular "laws" describing the regularities characteristic of Chinese society; and that integration was only possible through a detailed investigation and close knowledge of current conditions and Chinese history. . . .

In "On the New Stage" (1938), Mao had asserted that Marxism had to be regarded as a guide to action.* He returned to this theme frequently in the *zhengfeng*[2] documents, and it represented the major theme of his keynote speech originally entitled "Reform in Learning, the Party and Literature" (1942):

> Our comrades must understand that we do not study Marxism-Leninism because it is pleasing to the eye, or because it has some mystical value. . . . It is only extremely useful. . . . Marx, Lenin, and Stalin have repeatedly said, "Our doctrine is not dogma; it is a guide to action." . . . Theory and practice can be combined only if men of the Chinese Communist Party take the standpoints, concepts, and methods of Marxism-Leninism, apply them to China, and create a theory from conscientious research on the realities of the Chinese revolution and Chinese history.[†]

For Mao, Sinified Marxism represented the union of Marxism's universal laws and the particular "laws" of Chinese society. How did he perceive this ideology as a "guide to action"? It must be stressed that this ideological system did not contain within it the formulae for automatic and necessarily correct responses to the various political, economic or military contingencies which might arise in the course of revolution. The function of the ideology was to facilitate as accurate an interpretation of the historical context as was possible. This information would allow the revolutionary to take judicious action commensurate with the objective limitations of the situation as outlined by the ideology. The action could only be regarded as appropriate in its conception (rather than as necessarily correct), for there could be no formula for "correct" action implicit in the information provided. Having a clear and, it could be hoped, accurate picture of the historical situation would act as a guide to action by ruling out inappropriate

---

*Schram, *The Political Thought of Mao*, 171.
[2]zhengfeng: Rectification Movement
[†]Boyd Compton, trans., *Mao's China: Party Reform Documents, 1942–44* (Seattle: University of Washington Press, 1952), 21–22.

MAO ZEDONG'S "SINIFICATION OF MARXISM"

Universal Laws of Marxism
incorporating general historical predictions

Application

Particular "Laws"
describing characteristics of China as a particular historical situation

=

Sinified Marxism

Guide to action:
provides the information allowing appropriate (not necessarily correct) responses to be formulated

Action

Action and its effects become a feature of the specific historical context.

Evaulation:
adjustment of response through experience and practice

This chart maps the stages of Mao's approach to solving problems and making Marxism fit China.

responses and presenting certain actions as preferable, or perhaps obvious.

It is in this context that Mao's theory of practice finds relevance. Ideology could only serve as a guide to action by presenting an accurate assessment of the historical situation or process. It was up to the political actor to take full cognizance of the particular "laws" of the situation, to draw the necessary inferences and formulate an appropriate response accordingly. Such a response could not be regarded as "correct" in advance of its implementation, only as appropriate. The only method of ascertaining whether the seemingly appropriate action was correct was by performing the action and evaluating its results. If there was an equivalence between intention and result, then the action and the interpretation upon which it was based were indeed correct; otherwise the disparity between intention and result served to indicate either faulty analysis of the situation, or formulation of seemingly appropriate but incorrect response. Only by thus engaging reality could experience be gained and action refined so that the gap between the seemingly appropriate and the correct response could be minimised.

14

LI ZHISUI

The Emperor of Zhongnanhai
1994

Li Zhisui, Mao's private doctor on and off from the mid-1950s on, published in English this denunciation of the Chairman as evil emperor in 1994. Li depicts Mao as a corrupt king in Zhongnanhai, the leadership compound of the CCP in downtown Beijing. Like Snow's book (see Document 11), Li's has been translated into Chinese and circulates widely (albeit illegally) in China today. It has been criticized by the CCP for demonizing the Chairman and by scholars inside and outside China for its lack of balance (see Document 17). Yet Li paints a vivid picture, and

Li Zhisui, *The Private Life of Chairman Mao*, trans. Tai Hung-chao (New York: Random House, 1994), 272–75, 276–78.

most scholars trust his ability to report what he saw directly. This selection on the Great Leap Forward year of 1958 shows the rarefied atmosphere in which Mao traveled by that time—shielded from unpleasant realities by his staff and essentially duped by their efforts to please him. This picture helps us understand how Mao could have come to think that the unrealistic agricultural policies and backyard steel furnaces of the Great Leap Forward were working.

On September 10, 1958, Mao set out again, traveling by plane, train, and boat, to see for himself the vast changes taking place in the country. His popular adulation grew at every stop we made.

We flew first to Wuhan. Two of Mao's most enthusiastic admirers—"democratic personage" and Guomindang defector Zhang Zhizhong and Anhui's first party secretary Zeng Xisheng—visited him there. Mao thrilled Zhang Zhizhong by inviting him to come along on his inspection tour, and Zhang obliged by showering Mao with flattery. "The condition of the country is excellent indeed," he said to Mao. "The weather is favorable, the nation is at peace, and the people feel secure."

Zeng Xisheng, too, was courting Mao's favor. He wanted the Chairman to visit Anhui province. Zhang Zhizhong, a native of Anhui himself, joined Zeng in encouraging the visit. Mao agreed. We took a boat down the Yangtze to the city of Anqing, just on the border of Anhui, where first party secretary Zeng Xisheng escorted our party by car to Anhui's capital, Hefei. There, we witnessed new miracles. "Backyard steel furnaces" were the local specialty.

I saw the first such furnace—a makeshift brick and mortar affair, four or five meters high—in the courtyard of the offices of the Anhui provincial party committee. The fire was going full-blast, and inside were all sorts of household implements made of steel—pots, pans, doorknobs, and shovels—being melted down to produce what Zeng assured Mao was also steel. Zeng Xisheng picked up a hot nugget from the ground, plucked from the furnace only moments before, to show Mao the fruit of the mill, and nearby were samples of finished steel, indisputable evidence of the success of the backyard steel furnace. Mao had called upon the country to overtake Great Britain in steel production within fifteen years, by using methods that were quick and economical. Even now, I do not know where the idea of the backyard steel furnaces originated. But the logic was always clear: Why spend millions of dollars building modern steel plants when steel could be produced for almost nothing in courtyards and fields? The "indigenous," or "backyard," steel furnace was the result.

I was astounded. The furnace was taking basic household implements and transforming them into nuggets called steel, melting down knives into ingots that could be used to make other knives. I had no idea whether the ingots were of good-quality steel, but it did seem ridiculous to melt steel to produce steel, to destroy knives to make knives. The backyard steel furnaces were everywhere in Anhui, all producing the same rough-looking ingots.

Toward the end of the visit, Zhang Zhizhong proposed that Mao ride through the streets in an open car so the citizens of Hefei could see their great leader. In the summer of 1949, Mao had entered Beijing in an open jeep and the citizens lined the streets to welcome their liberation. In September 1956, during a visit by Indonesian president Sukarno, Mao had ridden in an open cavalcade. But he rarely appeared so openly before the masses. The Chairman's provincial visits were almost always secret, and security was tight. When he visited factories, his exchanges with workers were carefully controlled. Mao's face-to-face meetings were ordinarily confined to high-ranking party elite or leaders of the "democratic" parties. His twice-yearly appearances on the top of Tiananmen were not really exceptions. The crowds in the square were carefully chosen. The risk of appearing publicly before the masses was not only to Mao's security. The Chairman did not want to be accused of fostering his own cult of personality.

Mao believed that the masses needed a great leader and that the chance to see him could have an inspirational, potentially transformative, effect. But he needed the illusion that the demand for his leadership came spontaneously from the masses themselves. He would not be guilty of having actively promoted his own cult of personality. "Democratic personage" Zhang Zhizhong, sensitive to Mao's dilemma, was well suited to push Mao into the limelight. "You seem very concerned about the development of a personality cult," Zhang said to Mao.

Zhang argued, though, that Mao was the Lenin, not the Stalin, of China. Mao, like Lenin, had led the Communist party and the Chinese people to revolutionary victory, living on to lead in the construction of socialism, too. Unlike Lenin, who had died only eight years after the success of revolution, Mao would bless the people of China with his leadership for another thirty or forty years, they hoped. The difference between Mao and Stalin was that Stalin had promoted his own cult of personality. Mao had not. Mao, said Zhang, had a democratic style of leadership that stressed the "mass line" and avoided arbitrariness and dictatorship. "How can our country have a personality cult?" he wondered. "Progress is so fast, and the improvement in the life of the people so great that the masses spontaneously pour out their sincere, passionate feelings for you. Our people truly love their great leader. This is not a personality cult." Mao loved Zhang's flattery. The two were a perfect pair. The Chairman agreed to show himself to the citizens of Hefei.

On September 19, 1958, over 300,000 people lined the streets of Hefei hoping for a glimpse of Mao. He rode slowly through the city in an open car, waving impassively to the throngs, basking in their show of affection. I suspect that the crowds in Hefei were no more spontaneous than those in Tiananmen. The gaily colored clothes, the garlands of flowers around their necks, the bouquets they held aloft as the motorcade passed by, the singing, the dancing, the slogans they shouted—"Long Live Chairman Mao," "Long Live the People's Communes," "Long Live the Great Leap Forward"—suggested that Zeng Xisheng had left little to chance. These crowds had also been carefully chosen, directed by the Anhui bureau of public security. But the crowds were no less enthusiastic, no less sincere in their adulation, for having been carefully chosen. At the sight of their Chairman, they went wild with delight. . . .

. . . Mao's plan for the Great Leap Forward was grandiose, utopian— to catch up with Great Britain in fifteen years, to transform agricultural production, using people's communes to walk the road from socialism to communism, from poverty to abundance. Mao was accustomed to sycophancy and flattery. He had been pushing the top-level party and government leaders to embrace his grandiose schemes. Wanting to please Mao, fearing for their own political futures if they did not, the top-level officials put pressure on the lower ones, and lower-level cadres complied both by working the peasants relentlessly and by reporting what their superiors wanted to hear. Impossible, fantastical claims were being made. Claims of per-*mu* grain production went from 10,000 to 20,000 to 30,000 pounds.

Psychologists of mass behavior might have an explanation for what went wrong in China in the late summer of 1958. China was struck with a mass hysteria fed by Mao, who then fell victim himself. We returned to Beijing in time for the October first celebrations.[1] Mao

---

[1]October 1 is National Day in China, commemorating the founding of the People's Republic in 1949.

began believing the slogans, casting caution to the winds. Mini–steel mills were being set up even in Zhongnanhai, and at night the whole compound was a sea of red light. The idea had originated with the Central Bureau of Guards, but Mao did not oppose them, and soon everyone was stoking the fires—cadres, clerks, secretaries, doctors, nurses, and me. The rare voices of caution were being stilled. Everyone was hurrying to jump on the utopian bandwagon. Liu Shaoqi, Deng Xiaoping, Zhou Enlai, and Chen Yi,[2] men who might once have reined the Chairman in, were speaking with a single voice, and that voice was Mao's. What those men really thought, we never will know. Everyone was caught in the grip of this utopian hysteria.

Immediately after the October first celebrations, we set out again by train, heading south. The scene along the railroad tracks was incredible. Harvest time was approaching, and the crops were thriving. The fields were crowded with peasants at work, and they were all women and young girls dressed in reds and greens, gray-haired old men, or teenagers. All the able-bodied males, the real farmers of China, had been taken out of agricultural production to tend the backyard steel furnaces.

The backyard furnaces had transformed the rural landscape. They were everywhere, and we could see peasant men in a constant frenzy of activity, transporting fuel and raw materials, keeping the fires stoked. At night, the furnaces dotted the landscape as far as the eye could see, their fires lighting the skies.

Every commune we visited provided testimony to the abundance of the upcoming harvest. The statistics, for both grain and steel production, were astounding. "Good-news reporting stations" were being set up in communal dining halls, each station competing with nearby brigades and communes to report—red flags waving, gongs and drums sounding—the highest, most extravagant figures.

Mao's earlier skepticism had vanished. Common sense escaped him. He acted as though he believed the outrageous figures for agricultural production. The excitement was contagious. I was infected, too. Naturally, I could not help but wonder how rural China could be so quickly transformed. But I was seeing that transformation with my own eyes. I allowed myself only occasional, fleeting doubts.

One evening on the train, Lin Ke tried to set me straight. Chatting with Lin Ke and Wang Jingxian, looking out at the fires from the back-

---

[2]These were all senior leaders of the CCP.

yard furnaces that stretched all the way to the horizon, I shared the puzzlement I had been feeling, wondering out loud how the furnaces had appeared so suddenly and how the production figures could be so high.

What we were seeing from our windows, Lin Ke said, was staged, a huge multi-act nationwide Chinese opera performed especially for Mao. The party secretaries had ordered furnaces constructed everywhere along the rail route, stretching out for ten *li* on either side, and the women were dressed so colorfully, in reds and greens, because they had been ordered to dress that way. In Hubei, party secretary Wang Renzhong had ordered the peasants to remove rice plants from faraway fields and transplant them along Mao's route, to give the impression of a wildly abundant crop. The rice was planted so closely together that electric fans had to be set up around the fields to circulate air in order to prevent the plants from rotting. All of China was a stage, all the people performers in an extravaganza for Mao.

The production figures were false, Lin Ke said. No soil could produce twenty or thirty thousand pounds per *mu*. And what was coming out of the backyard steel furnaces was useless. The finished steel I had seen in Anhui that Zeng Xisheng claimed had been produced by the backyard steel furnace was fake, delivered there from a huge, modern factory.

"This isn't what the newspapers are saying," I protested.

The newspapers, too, were filled with falsehoods, Lin Ke insisted, printing only what they had been told. "They would not dare tell the public what was really happening," Lin said.

I was astonished. The *People's Daily* was our source of truth, the most authoritative of all the country's newspapers. If the *People's Daily* was printing falsehoods, which one would tell the truth?

15

RAE YANG

At the Center of the Storm
1997

The Red Guard student groups were at the heart of the Cultural Revolution. Mao paraded them in Tiananmen Square and they roamed with virtual impunity around the country in 1966 and 1967. On the one hand, the Red Guards were utterly idealistic in their desire to serve and follow Chairman Mao. On the other hand, this radical idealism gave vent to the petty resentments of students and frequently led to violent deaths. Rae Yang's recent memoir of her life as a Red Guard demonstrates both results. At first she found the Cultural Revolution thrilling. She created dazibao, or "big character posters"—big sheets of paper denouncing anyone or anything the Red Guards considered "counterrevolutionary." Her targets, however, were merely high school teachers she didn't like, not the national political figures criticized by Mao. She was transfixed when she caught a glimpse of Chairman Mao at Tiananmen Square. Later, when her Red Guard troop traveled to Guangzhou to "make revolution," they ended up beating a deranged man to death. What began with pure ideals ended with sordid death. The question to consider here is, How much of the complex ideas in Mao's writings did Rae Yang and her fellow Red Guards understand?

From May to December 1966, the first seven months of the Cultural Revolution left me with experiences I will never forget. Yet I forgot things almost overnight in that period. So many things were happening around me. The situation was changing so fast. I was too excited, too jubilant, too busy, too exhausted, too confused, too uncomfortable. . . . The forgotten things, however, did not all go away. Later some of them sneaked back into my memory, causing me unspeakable pain and shame. So I would say that those seven months were the most terrible in my life. Yet they were also the most

Rae Yang, *Spider Eaters: A Memoir* (Berkeley: University of California Press, 1997), 115–18, 122–23, 136–38.

wonderful! I had never felt so good about myself before, nor have I ever since.

In the beginning, the Cultural Revolution exhilarated me because suddenly I felt that I was allowed to think with my own head and say what was on my mind. In the past, the teachers at 101[1] had worked hard to make us intelligent, using the most difficult questions in mathematics, geometry, chemistry, and physics to challenge us. But the mental abilities we gained, we were not supposed to apply elsewhere. For instance, we were not allowed to question the teachers' conclusions. Students who did so would be criticized as "disrespectful and conceited," even if their opinions made perfect sense. Worse still was to disagree with the leaders. Leaders at various levels represented the Communist Party. Disagreeing with them could be interpreted as being against the Party, a crime punishable by labor reform, imprisonment, even death.

Thus the teachers created a contradiction. On the one hand, they wanted us to be smart, rational, and analytical. On the other hand, they forced us to be stupid, to be "the teachers' little lambs" and "the Party's obedient tools." By so doing, I think, they planted a sick tree; the bitter fruit would soon fall into their own mouths.

When the Cultural Revolution broke out in late May 1966, I felt like the legendary monkey Sun Wukong,[2] freed from the dungeon that had held him under a huge mountain for five hundred years. It was Chairman Mao who set us free by allowing us to rebel against authorities. As a student, the first authority I wanted to rebel against was Teacher Lin, our homeroom teacher—in Chinese, *banzhuren*. As *banzhuren*, she was in charge of our class. A big part of her duty was to make sure that we behaved and thought correctly. . . .

Now the time had come for the underdogs to speak up, to seek justice! Immediately I took up a brush pen, dipped it in black ink and wrote a long *dazibao* (criticism in big characters). Using some of the rhetorical devices Teacher Lin had taught us, I accused her of lacking proletarian feelings toward her students, of treating them as her enemies, of being high-handed, and suppressing different opinions. When I finished and showed it to my classmates, they supported me by signing their names to it. Next, we took the *dazibao* to Teacher Lin's home

---

[1] *101:* an elite high school in Beijing attached to Beijing University and generally reserved for the children of high party officials or famous scholars or artists

[2] *Sun Wukong:* the lead character in the classical and popular Chinese novel *Journey to the West* (also translated in an abridged version as *Monkey*). He is a naughty and powerful character and was a great favorite of Mao.

nearby and pasted it on the wall of her bedroom for her to read carefully day and night. This, of course, was not personal revenge. It was answering Chairman Mao's call to combat the revisionist educational line. If in the meantime it caused Teacher Lin a few sleepless nights, so be it! This revolution was meant to "touch the soul" of people, an unpopular teacher in particular.

Teacher Lin, although she was not a good teacher in my opinion, was not yet the worst. Teacher Qian was even worse. He was the political teacher who had implemented the Exposing Third Layer of Thoughts campaign.[3] In the past many students believed that he could read people's minds. Now a *dazibao* by a student gave us a clue as to how he acquired this eerie ability. Something I would not have guessed in a thousand years! He had been reading students' diaries in class breaks, while we were doing physical exercise on the sports ground. The student who wrote the *dazibao* felt sick one day and returned to his classroom earlier than expected. There he had actually seen Qian sneak a diary from a student's desk and read it. The student kept his silence until the Cultural Revolution, for Qian was his *banzhuren.*

So this was Qian's so-called "political and thought work"! What could it teach us but dishonesty and hypocrisy? Such a "glorious" example the school had set for us, and in the past we had revered him so much! Thinking of the nightmare he gave me, I was outraged. "Take up a pen, use it as a gun." I wrote another *dazibao* to denounce Teacher Qian.

Within a few days *dazibao* were popping up everywhere like bamboo shoots after a spring rain, written by students, teachers, administrators, workers, and librarians. Secrets dark and dirty were exposed. Every day we made shocking discoveries. The sacred halo around the teachers' heads that dated back two thousand five hundred years to the time of Confucius disappeared. Now teachers must drop their pretentious airs and learn a few things from their students. Parents would be taught by their kids instead of vice versa, as Chairman Mao pointed out. Government officials would have to wash their ears to listen to the ordinary people. Heaven and earth were turned upside down. The rebellious monkey with enormous power had gotten out. A revolution was underway. . . .

[3] *Exposing Third Layer of Thoughts campaign:* an earlier ideological remolding campaign in which Rae Yang and her classmates had participated

On August 18, 1966, I saw Chairman Mao for the first time. The night before, we set off from 101 on foot a little after midnight and arrived at Tiananmen Square before daybreak. In the dark we waited anxiously. Will Chairman Mao come? was the question in everybody's mind. Under a starry sky, we sang.

"Lifting our heads we see the stars of Beidou [the Big Dipper], lowering our heads we are longing for Mao Zedong, longing for Mao Zedong. ..."

We poured our emotions into the song. Chairman Mao who loved the people would surely hear it, for it came from the bottom of our hearts.

Perhaps he did. At five o'clock, before sunrise, like a miracle he walked out of Tiananmen onto the square and shook hands with people around him. The square turned into a jubilant ocean. Everybody was shouting "Long live Chairman Mao!" Around me girls were crying; boys were crying too. With hot tears streaming down my face, I could not see Chairman Mao clearly. He had ascended the rostrum. He was too high, or rather, the stands for Red Guard representatives were too low.

Earnestly we chanted: "We-want-to-see-Chair-man-Mao!" He heard us! He walked over to the corner of Tiananmen and waved at us. Now I could see him clearly. He was wearing a green army uniform and a red armband, just like all of us. My blood was boiling inside me. I jumped and shouted and cried in unison with a million people in the square. At that moment, I forgot myself; all barriers that existed between me and others broke down. I felt like a drop of water that finally joined the mighty raging ocean. I would never be lonely again.

[Later in the fall of 1966, Rae Yang and her Red Guard troop traveled south to Guanzhou to spread their revolution. Early in their stay, the following incident occurred.]

On that night, two female Red Guards who were senior students did not come back until after nine o'clock. We were beginning to worry about them. Then we saw them return with a "captive," who was a big, stout man in his thirties. They explained to us why they had "arrested" this man.

In the afternoon the two Red Guards got lost in the city. Because of the directions this man gave them, they ended up in an abandoned cathedral in the suburbs. In twilight the two young women wandered about the ruin, trying to figure out what went wrong and how to get back to the city. Around them the weeds were tall and the trees were casting long shadows. The wind rustled and insects chirped. Suddenly they heard a commotion behind them. It turned out that a group of local people had seized a man.

It was the same man who had given them the wrong directions; then he followed them all the way to the cathedral. The female Red Guards did not notice him, but the local people, whose revolutionary vigilance had been heightened, became suspicious. They knew that rape had been committed on this site.

Hearing this, I was shocked. Rape! In my mind, it was a crime almost as bad as murder. So we interrogated him. What he said about his name, age, and profession has escaped me completely. We must have inquired into his family background and class status too. Probably he did not belong to the Five Red Categories (workers, poor and lower-middle peasants, revolutionary cadres, revolutionary servicemen, revolutionary martyrs), or else what happened that night might not have happened. In my memory even his face is fuzzy, like a picture out of focus. The only thing I remember clearly is the pair of white cotton shorts he had on that night.

To our angry question of why he had tricked our two comrades into the deserted cathedral, he could not give a satisfactory explanation. That convinced us that he had harbored evil intentions toward our class sisters. We closed in on him. Hands on hips. Fingers pointed at the tip of his nose. Some were already unbuckling their belts. Our questions became sharp.

"So do you hate Red Guards? Tell us the truth! Or else we'll smash your dog head!"

"Yes. I hate the Red Guards."

"Then do you hate the Cultural Revolution too? Do you want to sabotage the Cultural Revolution?"

"Yes. Yes. I hate . . . I want to sabotage . . ."

"Are you a class enemy?"

"Yes. I am a class enemy."

"Are you a Nationalist agent?"

"Yes. I'm an agent. I came from Taiwan."

"Do you hope the Nationalists come back?"

"Yes. I do . . ."

"Do you have guns?"

"Oh yes. I have guns. I have grenades too. I even have a machine gun."

"And a transmitter-receiver to contact Taiwan?"

"Sure. I have a transmitter-receiver."

"Where did you hide these?"

"I buried them in my backyard. You come with me. I'll take you there. You can dig them out."

As the interrogation went on, the man confessed that he had committed all the crimes we could think of. The words that dropped out of his mouth turned into facts in our minds. And these "facts" fueled our hatred toward him. He was no longer a suspect. He had become a criminal, a real class enemy. We started to beat him.

The next thing he did was a real shock to all of us. In a shower of fists, kicks, curses, and thrashes, he suddenly straightened up and pulled his white cotton shorts down. He had no underwear on. So there was his thing, his penis. Large and black. It stuck out from a clump of black hair. To me it seemed erect, nodding its head at all of us.

I couldn't help staring at it. I was dumbfounded. I was embarrassed. I was furious. My hands were cold and my cheeks were on fire. For a few seconds none of us moved. We were petrified. Then the dike burst. Torrents of water rushed out. All the female Red Guards ran out of the classroom. We stayed in the corridor. The male Red Guards charged forward. On their way they picked up long bamboo sticks to hit him.

We all hated him! I could not tell who hated him more. The female Red Guards hated him because he had insulted all of us. The male Red Guards hated him too, because he was a scum of their sex. By exposing himself, he had exposed all of them. They were stripped. They were shamed. This time they beat him hard. No mercy on him. He did not deserve it. He was a bad egg!

The sticks fell like rain. In a few minutes, the man dropped to the ground. The sticks stood in midair. Then someone pulled his shorts back up. After that we streamed back into the classroom. We looked. He did not move. He did not breathe. This man was dead!

16

CENTRAL COMMITTEE OF THE CHINESE COMMUNIST PARTY

Some Questions on Party History
June 1981

After Mao's death in 1976, there was a period of political transition. By 1979 the new leadership under Deng Xiaoping had consolidated, and it turned its attention to righting the wrongs it had seen in the Cultural Revolution. This left the party with a delicate task: how to acknowledge Mao's errors in the Cultural Revolution without discrediting the entire CCP. The resolution presented here was passed by the Central Committee on June 27, 1981. Like its predecessor, the first resolution on party history passed at the Seventh Party Congress in Yan'an in 1945, this resolution sought to lay to rest unresolved issues and to buttress the current leadership. It was decidedly less successful than the 1945 version in achieving this goal.

We can see in this short extract that in 1981 the CCP wanted to stress the "collective wisdom" nature of "Mao Zedong Thought," along with the contributions of several other party leaders, named in the first paragraph. In addition, the resolution lays the blame for the Cultural Revolution on Mao but tries to separate Mao the man from Mao the ideological lifeblood of the party. The chief logic used is that because of the collaborative nature of "Mao Zedong Thought," with so many contributing to it, the failings of one man, even the figurehead, should not discredit the system.

[The text begins with a two-page summary of the achievements of party rule from 1956 to 1966.]

All the successes in these 10 years were achieved under the collective leadership of the Central Committee of the Party headed by Comrade Mao Zedong. Likewise, responsibility for the errors committed in the work of this period rested with the same collective leadership.

Translation selected from *Beijing Review*, 6 July 1981, 10–39.

Although Comrade Mao Zedong must be held chiefly responsible, we cannot lay the blame on him alone for all those errors. During this period, his theoretical and practical mistakes concerning class struggle in a socialist society became increasingly serious, his personal arbitrariness gradually undermined democratic centralism in Party life and the personality cult grew graver and graver. The Central Committee of the Party failed to rectify these mistakes in good time. Careerists like Lin Biao, Jiang Qing and Kang Sheng, harboring ulterior motives, made use of these errors and inflated them. This led to the inauguration of the "cultural revolution."....

Chief responsibility for the grave "Left" error of the "cultural revolution," an error comprehensive in magnitude and protracted in duration, does indeed lie with Comrade Mao Zedong. But after all it was the error of a great proletarian revolutionary. Comrade Mao Zedong paid constant attention to overcoming shortcomings in the life of the Party and state. In his later years, however, far from making a correct analysis of many problems, he confused right and wrong and the people with the enemy during the "cultural revolution." While making serious mistakes, he repeatedly urged the whole Party to study the works of Marx, Engels and Lenin conscientiously and imagined that his theory and practice were Marxist and that they were essential for the consolidation of the dictatorship of the proletariat. Herein lies his tragedy....

Comrade Mao Zedong's prestige reached a peak and he began to get arrogant at the very time when the Party was confronted with the new task of shifting the focus of its work to socialist construction, a task for which the utmost caution was required. He gradually divorced himself from practice and from the masses, acted more and more arbitrarily and subjectively, and increasingly put himself above the Central Committee of the Party. The result was a steady weakening and even undermining of the principle of collective leadership and democratic centralism in the political life of the Party and the country. This state of affairs took shape only gradually and the Central Committee of the Party should be held partly responsible. From the Marxist viewpoint, this complex phenomenon was the product of given historical conditions. Blaming this on only one person or on only a handful of people will not provide a deep lesson for the whole Party or enable it to find practical ways to change the situation....

Comrade Mao Zedong was a great Marxist and a great proletarian revolutionary, strategist and theorist. It is true that he made gross mistakes during the "cultural revolution," but, if we judge his activities as a whole, his contributions to the Chinese revolution far outweigh his mistakes. His merits are primary and his errors secondary. He rendered indelible meritorious service in founding and building up our Party and the Chinese People's Liberation Army, in winning victory for the cause of liberation of the Chinese people, in founding the People's Republic of China and in advancing our socialist cause. He made major contributions to the liberation of the oppressed nations of the world and to the progress of mankind.

The Chinese Communists, with Comrade Mao Zedong as their chief representative, made a theoretical synthesis of China's unique experience in its protracted revolution in accordance with the basic principles of Marxism-Leninism. This synthesis constituted a scientific system of guidelines befitting China's conditions, and it is this synthesis which is Mao Zedong Thought, the product of the integration of the universal principles of Marxism-Leninism with the concrete practice of the Chinese revolution. Making revolution in a large Eastern semi-colonial, semi-feudal country is bound to meet with many special, complicated problems, which cannot be solved by reciting the general principles of Marxism-Leninism or by copying foreign experience in every detail. The erroneous tendency of making Marxism a dogma and deifying Comintern resolutions and the experience of the Soviet Union prevailed in the international communist movement and in our Party mainly in the late 1920s and early 1930s, and this tendency pushed the Chinese revolution to the brink of total failure. It was in the course of combating this wrong tendency and making a profound summary of our historical experience in this respect that Mao Zedong Thought took shape and developed. It was systematized and extended in a variety of fields and reached maturity in the latter part of the Agrarian Revolutionary War and the War of Resistance Against Japan, and it was further developed during the War of Liberation and after the founding of the People's Republic of China. Mao Zedong Thought is Marxism-Leninism applied and developed in China; it constitutes a correct theory, a body of correct principles and a summary of the experiences that have been confirmed in the practice of the Chinese revolution, a crystallization of the collective wisdom of the Chinese Communist Party. Many outstanding leaders of our Party made important contributions to the formation and development of Mao Zedong Thought, and they are synthesized in the scientific works of Comrade Mao Zedong. . . .

Mao Zedong Thought is the valuable spiritual asset of our Party. It will be our guide to action for a long time to come. The Party leaders and the large group of cadres nurtured by Marxism-Leninism and Mao Zedong Thought were the backbone forces in winning great victories for our cause; they are and will remain our treasured mainstay in the cause of socialist modernization. While many of Comrade Mao Zedong's important works were written during the periods of new-democratic revolution and of socialist transformation, we must still constantly study them. This is not only because one cannot cut the past off from the present and failure to understand the past will hamper our understanding of present-day problems, but also because many of the basic theories, principles and scientific approaches set forth in these works are of universal significance and provide us with invaluable guidance now and will continue to do so in the future. Therefore, we must continue to uphold Mao Zedong Thought, study it in earnest and apply its stand, viewpoint, and method in studying the new situation and solving the new problems arising in the course of practice.

17

JEFFREY WASSERSTROM

Mao Matters

1996

Jeffrey Wasserstrom is a historian of Republican China and popular culture. He brings a broader prospective to this review of recent scholarship on Mao than a Mao specialist might. His review is useful because it considers two of the most important books on Mao from the 1990s—Li Zhisui's immensely popular The Private Life of Chairman Mao *(see Document 14) and the challenging postmodern analysis of Mao's political impact by David Apter and Tony Saich in* Revolutionary Discourse in Mao's Republic. *Wasserstrom gives us a sense of the major trend in academic Mao studies today—the shift from studying Mao as either isolated "great man" or symbol of the revolution to studying how and why hundreds of millions of Chinese believed so strongly in him.*

Jeffrey Wasserstrom, "Mao Masters: A Review Essay," *China Review International,* vol. 3, no. 1 (Spring 1996): 1–21.

The two books under review here[1] are obviously of very different kinds—so different, in fact, that some readers may find it odd to find them paired in a review essay. David Apter and Tony Saich are academically trained political scientists who have joined forces to create an often insightful but sometimes difficult-to-follow scholarly study based on the collection and analysis of a variety of source materials, ranging from famous writings by Mao to interviews with survivors of the crucial Yan'an period in the history of the Chinese Communist Party (CCP), and this study makes use of theories and methodologies associated with such fields as anthropology and literary criticism. The recently deceased Li Zhisui, on the other hand, was trained as a physician rather than an academic, and his book is a straightforward, accessible, and deeply personal narrative based almost exclusively on the author's memory of his own encounters with Mao and conversations with members of the Chairman's inner circle.

It should also be noted that each book focuses on a different time and place: Yan'an during the early 1940s in the first case, and Beijing between 1954 (the year Li began serving as Mao's doctor) and 1976 (the year the Chairman died) in the second. Even though these books—which for convenience I shall refer to below as *Mao's Republic* and *Private Life*—are by different kinds of authors, take different forms, and deal with different periods, there are two reasons why it makes sense to place them side by side. First, each in its own way is among the most important contributions to date to what might loosely be called "Mao studies." Second, each directs our attention to one of the two main, diverging directions in which people attempting to come to terms with the Chairman's life and legacy are currently being pulled.

This is because one of the themes that is stressed in *Mao's Republic* is that the man once referred to as the "Great Helmsman" needs to be understood as a multidimensional and extremely complex figure. Apter and Saich's theoretically sophisticated analysis of the rhetorical dimensions of Yan'an politics leaves us with a picture of Mao that is nuanced and shaded. At some points, the authors remind us that Mao tended to view certain ends as justifying even the most brutal of means, but in other places they insist that we see this same man as a prophet with an unusual talent for giving voice to collective dreams and aspirations. In still other instances, they stress Mao's significance

[1]David E. Apter and Tony Saich, *Revolutionary Discourse in Mao's Republic* (Cambridge: Harvard University Press, 1994); Li Zhisui, *The Private Life of Chairman Mao: The Memoirs of Mao's Personal Physician Dr. Li Zhisui*, trans. Tai Hung-chao (New York: Random House, 1994).

as both the central subject in and a skilled creator of powerful mythic tales about the heroic redemption of lost patrimonies—tales that would retain their potency long after the People's Republic of China (PRC) was founded in 1949.

By contrast, *Private Life* presents us with an essentially monochromatic image of Mao. According to Li Zhisui, if we strip away all of the mythical images that have surrounded his famous (or infamous) patient's life and theories, the man we are left with is nothing more than a tyrant. The Mao that Li describes is a man who was driven above all else by an unbridled lust for absolute power, an extreme paranoia that made him doubt the loyalty of even his most sycophantic allies, and an unquenchable desire to satisfy a variety of hedonistic urges. Official propaganda may have presented the Great Helmsman as infinitely compassionate, but according to *Private Life,* the Chairman was actually an unfeeling dictator who never showed the slightest concern for the misery suffered by others as a result of his policies.

[Wasserstrom reviews earlier Mao studies briefly and then turns to representative works published in the early reform period following Mao's death.]

One of the earliest and most impressive products of this optimistic period was *Mao Tse-tung in the Scales of History,* a collection of essays that appeared in 1977. Dick Wilson, then the editor of the *China Quarterly,* followed a simple plan in putting together this volume: he invited prominent scholars who had previously written pieces on Mao to contribute new essays that focused on specific aspects of the leader's personality and accomplishments. The result was an appealingly diverse and often illuminating work; it contained chapters with such simple titles as "The Soldier" (by Jacques Guillermaz) and "The Chinese" (by Wang Gungwu). Wilson aptly described this collection as a "provisional" assessment of a figure who remained so complex and enigmatic that no one person could be expected to write a "just epitaph."* Throughout the decade or so that followed its publication, the volume stood out as one of the most ambitious and effective efforts by Mao specialists to reveal the man behind the myths. It was greeted by many as an unusually successful effort to bring together in one place all the separate "Maos" that needed to concern us. . . .

---

*Dick Wilson, *Mao Tse-tung in the Scales of History* (Cambridge: Cambridge University Press, 1977), viii.

... Many contemporary readers of the almost twenty-year-old *Mao Tse-tung in the Scales of History* will react to it in one of two very different ways.* Some will wonder why so many important "Maos" were left out of that book, and ask why chapters with such titles as "The Icon," "The Commodity," "The God," and "The Contested Symbol" were not included. In addition, critics may ask why a sharper distinction was not made between the very different "Maos" of different eras. They may even wonder, as I have, whether we might do well in the future to follow the lead suggested by the literature on figures as diverse as Karl Marx and Elvis Presley, and start to differentiate between a more attractive "earlier" or "young" Mao and a less appealing "later" or "old" Mao....

MULTIPLE MAOS

I want to argue here that Apter and Saich's book offers one of the strongest and most important cases to date for a plethora of different, though obviously interrelated and overlapping, Maos. ...

Apter and Saich distinguish between these various Maos in several different ways. For example, they speak at times of the need to remember that there was both a "hard" and a "soft" Mao at Yan'an. The former was someone who would do anything to secure and maintain authority. He was, in addition, a loyal member of a Leninist organization whose other top leaders tended to be just as ruthless as he was in the pursuit of power. The "soft" Mao, on the other hand, was a religious figure of sorts whom Apter and Saich portray as having had an almost magical gift for weaving together stories of personal and national loss and redemption into inspiring master narratives—narratives that helped to bind together and lend cohesion to what was in other ways a very fragmented and factionalized Yan'an community. The authors of *Mao's Republic* also stress the need to remember that the Chairman presented himself to this community in various guises, some of which combined the "hard" and "soft" sides of his persona. For example, at times he cultivated the image of himself as a wanderer in the wilderness searching for philosophical truths and honing his skills as a warrior, while on other occasions he and his allies pro-

*Good general overviews of recent trends in Chinese and Western treatments of Mao are provided in Brantly Womack, "Mao Zedong Thoughts," *The China Quarterly*, no. 137 (March 1994): 159–67; and Thomas Scharping, "The Man, the Myth, the Message—New Trends in Mao Literature from China," *The China Quarterly*, no. 137 (March 1994): 168–79.

moted the idea that he was an armed prophet who was equally at home writing sacred texts and leading the faithful into battle. . . .

THE MONSTROUS MAO

Li Zhisui's biases and basic feelings concerning Mao are spelled out at the very beginning of *Private Life*. In the Preface, the physician notes that he once "revered" his famous patient (p. xix) but then came to view him as a despicable creature, a "dedicated philanderer" whose only concerns were to satisfy sexual urges and maintain "total power" (p. xx). . . . This theme of imperial influence is so prominent in *Private Life* that it seems safe to speculate that, had a new edition of *Mao Tsetung in the Scales of History* been commissioned before the doctor's death and had Li Zhisui been asked to contribute a chapter, he might well have chosen to call his essay "The Emperor."

One would guess from the Foreword to *Private Life,* on the other hand, that Andrew Nathan might find "The Dictator" a more appropriate title for such a chapter. This is because, in introducing Li's memoir, Nathan moves beyond Chinese imperial metaphors to place Mao in a rogues' gallery that includes everyone from Caligula to Hitler (as well as Stalin, of course). What unites these tyrants, he suggests, is that they all experienced and were warped by "the deranging effects of absolute power" and ended up descending "into a shadow world, where great visions become father to great crimes" (pp. vii and xiv). . . .

A number of Chinese dissidents have also made much use of imperial and dynastic metaphors in their harsh critiques of autocratic Maoist rule, though in doing so they approach the issue from a somewhat different standpoint and clearly have quite different motivations compared to Western commentators.* What all of these various texts have in common is that they encourage us to think of a single historical Mao. They imply or explicitly state that the various Mao myths created in the PRC should be dismissed as nothing more than a delusionary smoke screen that temporarily hid the monster's true form. This highlights two key differences between *Private Life* and *Mao's Republic*. First, as already noted, Apter and Saich emphasize the need to think of multiple Maos. Second they insist that myths are more than just lies.

---

*See, e.g., Ruan Ming, *Deng Xiaoping: Chronicle of an Empire* (Boulder, Colo.: Westview Press, 1994), 80, 171, 234, passim.

This reformulation of the problem of mythmaking by Apter and Saich, which draws upon the work of Lévi-Strauss[2] and a variety of other theorists, has several interesting implications for our understanding of Mao. For example, it leads the authors of *Mao's Republic* to take quite seriously the famous interview-based account of Mao's childhood that Edgar Snow presents in *Red Star Over China,* which they note was widely read inside as well as outside China and "persuaded many young people to go to Yan'an" (p. 354 n. 35). Instead of dismissing Snow as a dupe who accepted at face value a largely erroneous account of Mao's early life, they see this text as providing an important window into the narrative and discursive strategies of self-presentation that CCP leaders turned to in their quest to build up a richly endowed fund of symbolic capital. Snow's text, in their eyes, "helped Mao to become a surrogate figure for the China to become," reinforced the idea that he was someone who was "undeterred by any obstacle to revolutionary learning," and, most importantly, served both to humanize Mao and make him "larger than life" (p. 90). . . .

THE VALUE OF THE TWO WORKS

After my criticism of both books for various failings, many may be tempted not to read either of them, but in fact each has a good deal to offer students of modern China. Mao remains such an important figure that even flawed efforts to understand him are worth looking at carefully, providing there is some kind of payoff. In the case of Li Zhisui's memoir, the most rewarding feature is the wealth of detailed information it provides about a wide variety of little-known people and places associated with Mao's inner circle. Skeptical readers (like myself) may in many cases end up wondering just how accurate a description of a specific locale or event really is—not so much because we doubt the doctor's integrity but rather because we wonder about the ability to remember things so clearly after such a long passage of time. Even skeptical readers, however, are likely to come away from *Private Life* convinced that they know more than they did before about a variety of aspects of PRC society that are usually kept hidden from view. . . .

[2] *Lévi-Strauss:* Claude Lévi-Strauss, a founding theorist of structuralist anthropology who has looked in particular at unconscious patterns of behavior in social groups. Many historians have found his theories useful in explaining human behavior.

Apter and Saich provide a much needed countervailing force to the many current works that seek to focus our attention exclusively on a single, monstrous Mao. Whether or not we give credence to the most appalling information that Li Zhisui presents to us in his memoir, there should be no question now that one Mao worth keeping in mind at all times is the man who became corrupted by power and capable of great callousness and cruelty. What Apter and Saich remind us of so effectively, however, is that there are other Maos to remember as well, including the rebel who was able to inspire a great deal of devotion, first among a band of followers in a remote part of China and later among a much broader segment of the Chinese population. Not only does one need to come to terms with the despot, but one must also come to terms with the visionary author responsible for powerful texts that drew attention so effectively to the humiliations that the nation had suffered at the hands of foreign oppressors and the peasants at the hands of landlords, and who in some cases even dealt eloquently with the abuses that women had suffered at the hands of husbands, fathers, lineage elders, and mothers-in-law. With a revolutionary leader who has been deified in as many ways as Mao has, it is certainly important to highlight his failings. But Mao is too important and intriguing a figure to reduce once again to a caricature. Those of us seeking to understand the confusing trajectories of the Chinese Revolution need to learn to think and talk in new ways about a multiplicity of contradictory Maos, ranging from the inspirational to the profoundly disturbing, and Apter and Saich have done us a valuable service by giving us a strong push in that direction.

18

GEREMIE BARMÉ

Shades of Mao

1990s

While the Chinese Communist party tries to limit Mao's aura and intellectuals use him to push their reform agendas, the ordinary people of China also have space in their hearts—and in their temples—for Chairman Mao. Geremie Barmé, the doyen of China pop and unofficial politics, opens the window to the devotional uses of Mao, as well as the irreverent, with two examples from his book Shades of Mao. *The first selection, by Xin Yuan, is an assessment from the Hong Kong press of Mao's role as a virtual deity in China's enduring popular religious traditions. The second selection is a punning rhyme that reflects Mao's meaning to today's working poor in China.*

A PLACE IN THE PANTHEON: MAO AND FOLK RELIGION BY XIN YUAN

(Published under a pen name, the Beijing scholar Wang Yi has written on such diverse subjects as traditional Chinese gardens and elements of folk religion in the political culture of the Cultural Revolution. This article appeared in the Hong Kong press at the end of 1992.)

From late last year [1991], China has experienced a craze that has involved the re-deification of Mao Zedong. It started in the South and has spread to the North. People have combined Mao's image with gold cash inscribed with the words "May This Attract Wealth"* or images of the eight hexagrams, and placed them in prominent places. Drivers throughout the country have Mao's picture hanging from their rear-view mirrors and claim that Mao can prevent car accidents. The cassette tape *The Red Sun: Odes to Mao Zedong Sung to a New*

*In Chinese, *zhaocai jinbao.*

Xin Yuan, "A Place in the Pantheon: Mao and Folk Religion," and "Musical Chairman," both from Geremie R. Barmé, *Shades of Mao: The Posthumous Cult of the Great Leader* (Armonk, N.Y.: M. E. Sharpe, 1996), 195–200, 283–84.

Beat has been a national best-seller. Mainlanders have variously called this the "Red Sun Phenomenon" or the "Mao Becomes a God Phenomenon." It has also given rise to numerous interpretations among political and cultural analysts.

Political conservatives are trying to dragoon this MaoCraze into the service of their efforts to "oppose peaceful evolution."* Deng Liqun has remarked "with the unprecedented international wave of revisionist thought coupled with the tide of Bourgeois Liberalism in China, we have indeed seen 'a miasmal mist once more rising.'"† But Deng has nothing to say on the subject of why the masses are now treating Mao Zedong like Zhao Gong or the Kitchen God.‡

Intellectuals, on the other hand, see the MaoCraze as evidence that elements of the Cultural Revolution still hold sway and that there's been no rational attempt to understand the long-term damage wrought by that period on the people of China. But such views are all superficial and simplistic. No one has tried to discuss the question in terms of the psychology of folk religion. Therefore, it is obvious that further analysis of the phenomenon is necessary.

The present Mao cult is different from the past in that it constitutes a popular deification of Mao, not a politically orchestrated one. People now seek the protection of the Mao-God when they build houses, engage in business, and drive vehicles. Old ladies place images of Mao over their stoves and in niches built for statues of the Buddha and burn incense to him morning and night. Traditional folk religion provides the real basis for the present Red Sun Craze.

In Chinese culture, the power of the gods is always reliant upon the authority of the ruler. As early as the *Zuozhuan,* in the Record of the Fourteenth Year of Duke Xiang, we find the unequivocal statement: "The ruler is the host of the spirits and the hope of the people."§ Politics and religion formed a mutually cooperative whole or, as [the late-Qing politician and military leader] Zeng Guofan put it, "The way of

*"Peaceful evolution" *(helping yanbian)* was officially regarded by the Chinese authorities as being the greatest threat to the communist system. It was supposedly a Western strategy that relied on peaceful rather than violent means—cultural, political and economic—to undermine communism and eventually replace it with a Western-style, free-market democratic government.

†See Deng Liqun, "Permanently On Heat," in *Shades of Mao.*

‡Zhao Gong (Zhao Xuantan, or *Zhao Gong yuanshuai*), the God of Wealth. The Kitchen God *(Zao jun, Zaoshen* or *Zaowang),* the spirit who rules over the hearth, was said to report back to Heaven every Chinese New Year's Eve.

§The *Zuozhuan* is a Confucian historical text of considerable antiquity. For this quotation, see James Legge, *The Chinese Classics (The Chu'un Ts'ew with The Tso Chuen),* p. 466, col. 2, and p. 462 for the Chinese original.

the kings rules in this world; the way of the gods in the other world."
Both ways witnessed a plethora of rulers, however, with dynasties rising and falling in the human world and, in the other world, rulers like
the Jade Emperor, Maitreya Buddha, and so on, gaining ascendancy at
one time or another. The only thing that did not change was the
immutable link between politics and religion.

With the communist takeover in 1949, popular religion in China
underwent the most violent change in its history. First, in the 1950s,
there was a movement to wipe out superstition, which was followed in
the 1960s by the call to "eliminate the Four Olds"* and the suppression of virtually all forms of religious activity in the country. The
effects were particularly devastating as both movements took place in
tandem with the creation of grass-roots Party cells and nationwide
Thought Reform. In traditional society, at the county level and lower,
political life was ruled by popular clan bodies that also had responsibility for other activities including religious observances. The post-1949
organization of society, however, saw this traditional arrangement
uprooted and the monopoly rule of Party committees at every level.
Folk religion was deprived entirely of the social and organizational
basis for its activities. Nonetheless, habits and practices that have
weathered changes over the millennia and provided spiritual succor
for people for so long are not so easily obliterated. Frustrated in its
traditional form, it is only natural that popular religious sentiment
would find new ways to express itself. In post-1949 China, the only
sanctified form it could take was in worshipping the Red Sun, Mao
Zedong.

The legacy of this intermingling of political and religious life is that
the people tend to view divine providence and spiritual power in political terms. . . . Harsh rulers were treated with awe and commemorated
in special temples with religious observances by later generations.
Because of this venerable tradition Mao's actual sentiment for the
people, or his munificence, or even his tyranny that was expressed so
succinctly in his line that "for the 800 million Chinese, struggling is a
way of life," is not really a major issue.† With the consolidation and
expansion of his power it was inevitable that he become deified.

---

*The Four Olds (sijiu) were: old ideas, old culture, old customs, and old habits, the
elimination of which was formalized in the Party's 16 Articles on the Cultural Revolution
promulgated in 1966.
†This Mao quote was published in a People's Daily editorial marking the tenth
anniversary of the Cultural Revolution.

Mao's deification was synchronous with his political apotheosis. According to reports in the *Beijing Evening News*, Mao badges made an appearance in and around Yan'an as early as 1945. After 1949, Mao's transmogrification[1] continued apace. . . .

Faith in the omnipresence of nonworldly power was reinforced during the Cultural Revolution. When everything else—all belief systems and cultural norms—was swept away and overthrown, Mao Zedong became the supreme and all-powerful super god, the "Sun that never sets." Mao was invested with a type of power equal, if not superior, to all other religious systems, expressed in such beliefs that he was the Sun, "sustainer of all things," a being who "turned the universe red." Because he was both omnipotent and omnipresent, people felt they could invest themselves with an "ever victorious" power through quasi-religious practices not dissimilar to shamanistic ritual and self-flagellation. They therefore paid homage to his image, sang Mao quotation songs, chanted his sayings, performed the Loyalty Dance, "struggled against self-interest and repudiated revisionism,"* and so on.

Faith in the power of the Red Sun in the Cultural Revolution was very much like ancient shamanistic belief. Both held that the power of the spirit could exorcise evil. In the past, people thought they were surrounded by malevolent forces that had to be subdued or expunged. Similarly, in the Cultural Revolution, there was a general belief that the world was full of evil subsumed under such rubrics as Imperialism and Revisionism, the Five Black Categories, as well as the grab-all expression Cow Demons and Snake Spirits.[†] People lived in constant fear that there could be "a restoration of capitalism," or that they would "have to suffer the bitterness of the past again," or that "millions would die" because of a disastrous counterrevolution. Since the Red Sun was a "Spiritual Atom Bomb" that could dispel all those evils, there is little wonder that the level of popular adoration and worship was so hysterical. The creation of a Red Sea (the ubiquitous displays

---

[1]*transmogrification:* grotesque or humorous change

*The Loyalty Dance *(zhongzi wu)* was a clumsy choreographic group dance designed to add rhythm to the adulation of Mao. To "struggle against self-interest and repudiate revisionism" *(dousi pixiu)* was a popular mantra that summed up the avowed aims of the Cultural Revolution.

†The Five Black Categories *(hei wulei)* were: landlords, rich peasants, counterrevolutionaries, bad elements, and rightists. Cow Demons and Snake Spirits *(niugui sheshen)* was a classical Chinese expression reinterpreted for the denunciation of people in the early stages of the Cultural Revolution.

of Mao's portrait, his quotations, slogans, and images that represented devotion to him) was a direct result of the "sweeping away of all Cow Demons and Snake Spirits" and "the rebellion against Class Enemies" that had seen the exorcising of evil in the first place. More recently, Deng Liqun's attempts to use the new MaoCraze as a weapon to repel the "miasmal mist once more rising" is little more than a continuation of ancient shamanistic practice.

. . . Traditional deities live on and this is why we see people throughout the country, in particular in the countryside, erecting temples to the God of Wealth, Door Gods, Guan Gong, Boddhisattvas, the King of Death, and other ancient icons to cure illness, for the begetting of male children, for help in making money, finding the right marriage partner, and for scholastic achievement.

The speed with which temples are being restored or built is comparable to the rate at which they were closed and destroyed in the 1950s and 1960s. According to the television documentary, *A Record of Modern Superstitions: Incense Burning*, broadcast on Beijing TV on 9 February 1990, more than a million people travel to the Southern Peak* to burn incense each year and spend some 100 million yuan on incense alone. The *Guangming Daily* has also reported that in recent years many shamans have appeared in Fang County, Hubei Province. Some of these are state cadres or retired cadres. Similar reports abound. Things are more controlled in the cities where years of centralized education have meant that primitive beliefs have to find new forms for expression. Because Mao has been the only all-powerful figure for so long, he was the obvious choice for popular adulation during the recent religious revival in China. He has become the idol to which the revived worship of the God of Wealth, Guan Gong, Guanyin Boddhisattva, and other gods is married.

It would seem farcical that Mao, a man who led the assault on capitalism during his lifetime, should in death be put on a par with the God of Wealth and inscribed with traditional imagery. But the misprision[2] and distortion of gods, their reinvention and reinterpretation, is a central element of popular religious activity. As was said long ago: "In a temple that has stood for five generations you can find all manner of strange things." The [Song dynasty literatus] Ouyang Xiu noted nearly a thousand years ago that people often misunderstand the spirits they

*The Southern Peak *(Nanyue)*, is Hengshan in Hunan Province, one of China's Five Sacred Mountains which feature in popular religious belief.
[2]*misprision:* act of contempt against a superior

worship when he wrote: "Of all distortions in the world those that one finds in a temple are the most extreme."

MUSICAL CHAIRMAN

The floating population, deracinated rural workers and beggars, *mangliu* in Chinese, became an increasingly major social problem as the economic Reforms continued in the 1980s and 1990s. Urban dwellers often spoke of the masses of roving peasants (tens to hundreds of millions depending on what sources you accepted) as being a major threat to social stability and future prosperity. Some *mangliu*, however, believed that it was from their ranks that a new strongman, someone perhaps with the stature of Mao Zedong, would eventually appear to rule the nation.

Mao himself had an early career as a *mangliu* of sorts, details of which are recorded in a book by his companion at the time, Siao-yu.* In the 1990s China's floating population armed itself with the invincible weapon of Mao Thought, or as a popular rhyming saying put it:

Beijing kao zhongyang,
Shanghai kao laoxiang,
Guangzhou kao Xianggang,
Mangliuzi kaode shi Mao Zedong sixiang.

北京靠中央，
上海靠老鄉，
廣州靠香港，
盲流子靠的是毛澤東思想。

Beijing relies on the Center,
Shanghai on its connections,[†]
Guangzhou leans on Hong Kong,
The drifting population lives by Mao Zedong Thought.[‡]

---

*Siao-yu, *Mao Tse-tung and I Were Beggars. A Personal Memoir of the Early Years of Chairman Mao.*

[†]Beijing relies on the power of Party Central to protect it and prosper; Shanghai on all the officials in Beijing of Shanghai provenance (the so-called "Shanghai Gang": Jiang Zemin, Zhu Rongji, Huang Ju, and so on).

[‡]In "Sailing the Seas Depends on the Helmsman" *(Dahai hangxing kao duoshou)*, the unofficial anthem of the Cultural Revolution, there is a line that goes: "The Revolutionary Masses rely on Mao Zedong Thought" *(geming qunzhong kaode shi Mao Zedong sixiang)*.

A China Chronology (1893–1976)

1893 Mao born to a farming family in Hunan province.

1895 China defeated by Japan in the Sino-Japanese War.

1900 Boxer Rebellion and counterattack by foreign powers.

1905 Qing dynasty ends the civil service exams.

1911 Republican revolution; fall of the Qing dynasty.

1912 General Yuan Shikai replaces Sun Yat-sen as president of the republic.

1913 Sun Yat-sen founds the Guomindang (GMD), or Nationalist party.

1915–
1925 New Culture Movement.

1919 May Fourth Movement in Beijing opposes the Versailles treaty.

1921 Official founding of the Chinese Communist party (CCP) in Shanghai; Mao Zedong is one of the founding members.

1923 CCP and GMD cooperate in the first united front; Mao then works in the GMD Propaganda Department and the Peasant Training Institute.

1925 Sun Yat-sen dies; Chiang Kai-shek eventually takes over the GMD.

1926 Northern Expedition under GMD leadership starts out from Guangzhou in southern China to reunify the country. Mao returns to Hunan to work in the countryside.

1927 Mao writes his "Report on the Peasant Movement in Hunan." Chiang Kai-shek turns against the CCP and in April massacres CCP activists.

GMD forms a new national government and moves the capital of the Republic of China to Nanjing.

1930 Changsha uprising, led by Mao, fails; Mao moves to develop rural soviets.

1931 Japanese begin to occupy northeast China and will declare Manchukuo a separate country the next year.

Formal establishment of the Jiangxi soviet; Mao elected chairman of the soviet.

1931–
1934 GMD national government launches five "encirclement campaigns" to destroy the Communist party and Jiangxi soviet; CCP survives the first four.

1934 Communists flee the Jiangxi soviet and set off on the Long March.

1935 Mao gains a central leadership position in the CCP at the Zunyi Conference during the Long March.

1936 Communists make their new capital at Yan'an, in the northwest province of Shaanxi.

1936–
1947 Yan'an period, during which Mao consolidates his supreme position and develops policies that lead to CCP victory in 1949.

1937 Japan invades China, beginning World War II in Asia.

1942–
1944 Yan'an Rectification Movement promotes Mao's ideas and leadership.

1945 Japan formally surrenders to the Allies on September 9, ending World War II.

1945–
1949 Chinese civil war between the CCP and GMD.

1949 CCP declares a new national government, the People's Republic of China (PRC), with Mao as its head.

1950 **February** China and the Soviet Union sign the Treaty of Friendship, Alliance, and Mutual Assistance.

October Chinese forces enter Korea, joining the Korean War.

1952 Land reform completed in most areas of China.

1955 First Five-Year Plan (1953–57) formally adopted to organize China's planned economy.

Nationwide mass campaign to criticize two intellectuals, Hu Shi and Hu Feng.

1956 Khrushchev denounces Stalin in the USSR.

Mao calls for public criticism of the CCP in the Hundred Flowers Campaign, but the response is muted.

CCP holds its Eighth Party Congress in Beijing; celebrates successes of the PRC.

1957 **February** Mao delivers his speech "On the Correct Handling of Contradictions among the People" to boost the Hundred Flowers Campaign.

May Hundred Flowers Campaign blooms, but with biting criticism of the CCP.

June Anti-Rightist Campaign attacks those who spoke up in the Hundred Flowers Campaign; Mao's "Contradictions" speech is published in a highly edited form.

1958 CCP adopts Mao's Great Leap Forward plan.

August Mao's talks at Beidaihe popularize collectivization of agriculture in the people's communes.

1959 Uprising in Tibet against Chinese rule is suppressed by the People's Liberation Army (PLA).

April Liu Shaoqi succeeds Mao as state chairman (Mao remains as party chairman).

August Peng Dehuai criticizes the Great Leap Forward and is purged, showing that Mao can no longer be criticized even by his senior colleagues.

1960 Soviet Union withdraws all experts from China.

Famine deepens in China.

1961 CCP begins economic and political reforms to undo the damage of the Great Leap Forward.

1964 China explodes its atomic bomb.

1966 Cultural Revolution begins as an attack on Beijing party intellectuals but quickly spreads to the purge of senior party leaders, including, by 1967, Liu Shaoqi.

Mao writes a big-character poster, "Bombard the Headquarters," to encourage Red Guards to attack "the four olds" and "to rebel is justified"; Red Guard terror begins.

1969 Mao declares the "victory" (that is, end) of the Cultural Revolution and supports General Lin Biao as his new successor; radical policies continue.

1971 **July** U.S. secretary of state Henry Kissinger secretly visits Beijing.

September Lin Biao dies in a plane crash while trying to escape Mao's secret police; he is formally denounced.

October China joins the United Nations, leading to the expulsion of Taiwan.

1972 U.S. president Richard Nixon visits China.

1976 **January** Zhou Enlai, the highest ranking moderate in the CCP, dies.

September Mao Zedong dies.

Questions for Consideration

1. How would you sum up Mao Zedong's life? Try making a list of dichotomies: Mao as hero vs. Mao as demon, Mao as state founder vs. Mao as voice of the people, Mao as thinker vs. Mao as actor, or Mao as historical figure vs. Mao as living icon. What other contrasts can you find?

2. Consider the different objects of Mao's attention in the three periods covered in our selections: social movements in the 1920s, state power and nationalism in the 1940s, and problems of administration and development in the 1950s and 1960s. Why did Mao's focus shift? What continuities do you find in Mao's writings over these years?

3. In what ways do the concrete policy suggestions and ways of looking at the world suggested in Mao's major writings before 1949 address the issues of imperialism and how to deal with the West— issues that confronted China so forcefully by 1900?

4. What experiences informed Mao's approach to rural revolution? Compare the picture Mao gave Snow (see Document 11) with other accounts of Mao's early life.

5. Why did Mao take the oppression of women and youths seriously? How did their liberation fit into his model of revolution? What became of Mao's "feminism"? Did the CCP follow his lead on women's and youth liberation over the years? Did Mao himself return to this theme in his later writings?

6. Why were the ideals for China outlined by Mao in "On New Democracy" (see Document 2) attractive to China's students and urban middle class in the 1940s?

7. Who were Mao's competitors for top leadership in the CCP in the 1940s, and how did his Yan'an writings (see Documents 2–4) answer their challenge? How did these writings function in the 1942–44 Rectification Movement?

8. How does Mao's view on revolutionary leadership (see Document 4) relate to his philosophical method, as explained in Document 13?

9. What issues confronted the ruling CCP in 1956? How did Mao try to address them in "On the Correct Handling of Contradictions among the People" (Document 7)? Why was Mao out of synch with his senior colleagues in the CCP?

10. How do Mao's Beidaihe talks of 1958 (see Document 8) compare with his Yan'an talks (see Documents 2–4)? Look at the contextual readings, such as Li Zhisui in Document 14 and Stuart Schram in Document 12, as you consider what contributed to these differences.

11. Why is it so difficult to separate Mao from the CCP today? Consider, for example, the party resolution of 1981 in Document 16 and popular views in the 1990s in Document 18.

12. How does the Mao cult, especially in the Cultural Revolution (see Documents 10 and 15) compare and contrast with other notable leadership cults of the twentieth century, such as Stalin's, Hitler's, or Mussolini's?

13. Land reform, or "land to the tiller," was a key issue for developmental states in the twentieth century. How did CCP policy and practice change from the 1940s to the 1960s to the 1980s?

14. Developmental states need to find a way to link an elite that can function in an international environment defined by Western science, legal norms, and commercial practices with a majority agrarian population that often holds radically different attitudes, values, and practices that are contrary to those international norms. How did Mao and the CCP try to bridge this gap?

15. Developmental states must find a way to achieve security—both international respect and domestic order. What were Mao's answers to these fundamental state questions?

Selected Bibliography

This bibliography gives a taste of the scholarship on Mao and his context.

MOVIES AND DRAMA

The images of films (many now available on video) and modern plays are an effective way to present the history of Mao's times and to give a sense of how various Chinese experienced Mao's revolutions. The movie *Huozhe* (To Live) by Zhang Yimo is a vivid account of a commoner's life through these years. It is readily available on video (with subtitles). Chen Kaige's *Farewell My Concubine* provides a parallel history from the perspective of the Chinese intellectual and artistic elite. *Small Well Lane*, a play written by Li Longyun and translated by Hong Jiang and Timothy Cheek (Ann Arbor: University of Michigan Press, 2002), is the compelling story of how people in a small Beijing neighborhood experienced Mao's revolutions from the 1940s to the 1980s.

BIBLIOGRAPHIES AND GUIDES

The best bibliography for those new to China studies is Charles Hayford, *China* (Santa Barbara, Calif.: Clio Press, 1997). There are several other fine bibliographies, such as the *Bibliography of Asian Studies*, available at research and college libraries.

The only English index of topics and names in Mao's works is John Bryan Starr and Nancy Anne Dyer, *Post-Liberation Works of Mao Zedong: A Bibliography and Index* (Berkeley: Center for Chinese Studies, University of California, 1979). However, the indexes in the comprehensive series of Mao's works in English by Schram and by Kau and Leung (see "Mao's Writings") are also useful.

Handy historical dictionaries with short entries on names, terms, and events in China's revolutions include the following:

He, Henry Yuhuai, *Dictionary of the Political Thought of the People's Republic of China* (Armonk, N.Y.: M. E. Sharpe, 2001).

Leung, Edwin Pak-wah, ed., *Historical Dictionary of Revolutionary China, 1839–1976* (Westport, Conn.: Greenwood Press, 1992).
Schoppa, R. Keith, *The Columbia Guide to Modern Chinese History* (New York: Columbia University Press, 2000).
Sullivan, Lawrence R., with Nancy Hearst, eds., *Historical Dictionary of the People's Republic of China: 1949–1997* (Lanham, Md.: Scarecrow Press, 1997).

TEXTBOOKS

Fairbank, John King, and Merle Goldman, *China: A New History* (Cambridge: Harvard University Press, 1998).
Lieberthal, Kenneth, *Governing China: From Revolution through Reform* (New York: W. W. Norton, 1995).
Saich, Tony, *Governance and Politics of China* (London and New York: Palgrave, 2001).
Spence, Jonathan, *The Search for Modern China*, 2nd ed. (New York: W. W. Norton, 2000).

MAO'S WRITINGS

The standard edition from China is *Selected Works*. There are now major scholarly editions of all periods of Mao's writings in English, which also include guides to the Chinese sources and interpretive issues.

Kau, Michael Y. M., and John K. Leung, eds., *The Writings of Mao Zedong, 1949–1976*, 2 vols. (Armonk, N.Y.: M. E. Sharpe, 1986, 1992).
MacFarquhar, Roderick, Timothy Cheek, and Eugene Wu, eds., *The Secret Speeches of Chairman Mao: From the Hundred Flowers to the Great Leap Forward* (Cambridge: Harvard Contemporary China Series, 1989).
Schram, Stuart, *The Political Thought of Mao Tse-tung*, rev. ed. (New York: Praeger, 1969).
Schram, Stuart, *Mao Tse-tung Unrehearsed: Talks and Letters 1956–1971* (Harmondsworth: Penguin, 1974).
Schram, Stuart R., and Nancy J. Hodes, eds., *Mao's Road to Power: Revolutionary Writings, 1912–1949* (Armonk, N.Y.: M. E. Sharpe, 1992–).
Selected Works of Mao Tse-tung (Peking: Foreign Languages Press, vols. 1–4, 1975; vol. 5, 1977). Available on-line at <http://www.marx2mao.org/Mao/Index.html>.

BIOGRAPHIES AND STUDIES OF MAO

There are numerous studies of Mao, but these are among the best. Of the three fine short biographies (Breslin, Davis, and Spence), Spence's is the most engaging. Schram's classic study, *Mao Tse-tung*, still repays reading. There are also a number of useful specialized studies. This section ends with some of the more thoughtful scholarly essays that assess these and other studies and put them in the context of scholarly opinion.

Breslin, Shaun, *Mao: Profiles in Power* (London: Longman, 1998).

Davin, Delia, *Mao Zedong* (Phoenix Mill: Sutton Publishing, 1997).

Dirlik, Arif, Paul Healy, and Nick Knight, *Critical Perspectives on Mao Zedong's Thought* (Atlantic Highlands, N.J.: Humanities Press, 1997).

Martin, Helmut, *Cult & Canon: The Origins and Development of State Maoism* (Armonk, N.Y.: M. E. Sharpe, 1982).

Schram, Stuart, *Mao Tse-tung*, rev. ed. (Harmondsworth: Penguin, 1967).

Schram, Stuart, *The Thought of Mao Tse-tung* (Cambridge: Cambridge University Press, 1989).

Spence, Jonathan, *Mao Zedong* (New York: Viking Penguin, 1999).

Starr, John Bryan, *Continuing the Revolution: The Political Thought of Mao* (Princeton: Princeton University Press, 1979).

Wakeman, Frederic, Jr., *History and Will: Philosophical Perspectives on Mao Tse-tung's Thought* (Berkeley: University of California Press, 1973).

Womack, Brantly, *The Foundations of Mao Tse-tung's Political Thought* (Honolulu: University of Hawaii Press, 1982).

Knight, Nick, "Mao and History: Who Judges and How?" *Australian Journal of Chinese Affairs*, no. 13 (January 1985): 121–36.

Scharping, Thomas, "The Man, the Myth, the Message—New Trends in Mao-Literature from China," *The China Quarterly*, no. 137 (1994): 168–79.

Schram, Stuart, "Mao Zedong a Hundred Years On: The Legacy of a Ruler," *The China Quarterly*, no. 137 (1994): 125–43.

Wasserstrom, Jeffrey N., "Mao Matters: A Review Essay," *China Review International*, vol. 3, no. 1 (1996): 1–21.

Womack, Brantly, "Mao Zedong Thoughts," *The China Quarterly*, no. 137 (1994): 159–67.

Womack, Brantly, "Mao before Maoism," *The Chinese Journal*, no. 46 (July 2001): 95–117.

CHINA'S REVOLUTIONS AND MAO

Long-Term Perspectives

Mao's contributions and faults can most fully be appreciated with an awareness of some long-term perspectives in Chinese history, as well as a sense of how Americans report on China. These are some challenging studies.

Jespersen, T. Christopher, *American Images of China: 1931–1949* (Stanford: Stanford University Press, 1996).

Levinson, Joseph, *Confucian China and Its Modern Fate* (Berkeley: University of California Press, 1968).

MacKinnon, Stephen, and Oris Friesen, *China Reporting: An Oral History of American Journalism in the 1930s and 1940s* (Berkeley: University of California Press, 1987).

Pussey, James, *China and Charles Darwin* (Cambridge: Harvard Council on East Asian Studies, 1983).

Schwartz, Benjamin I., *In Search of Wealth and Power: Yen Fu and the West* (Cambridge: Harvard University Press, 1964).

Wong, R. Bin, *China Transformed: Historical Change and the Limits of European Experience* (Ithaca: Cornell University Press, 1997).

Chinese Communist Party History

These fine studies and documentary collections give a sense of when and why Mao was in synch and out of synch with the Communist party (if not the broader Chinese population).

Guillermaz, Jacques, *A History of the Chinese Communist Party 1922–49* (New York: Random House, 1972).

Hartford, Kathleen, and Steven M. Goldstein, *Single Sparks: China's Rural Revolutions* (Armonk, N.Y.: M. E. Sharpe, 1989).

Hunt, Michael H., *The Genesis of Chinese Communist Foreign Policy* (New York: Columbia University Press, 1996).

Lampton, David M., *Same Bed Different Dreams: Managing U.S.-China Relations* (Berkeley: University of California Press, 2001).

Roy, Denny, *China's Foreign Relations* (Lanham, Md.: Rowman & Littlefield, 1998).

Saich, Tony, *The Rise to Power of the Chinese Communist Party: Documents and Analysis* (Armonk, N.Y.: M. E. Sharpe, 1996).

Teiwes, Frederick C., *Politics and Purges in China: Rectification and the Decline of Party Norms, 1950–1965*, 2nd ed. (Armonk, N.Y.: M. E. Sharpe, 1993).

1920s–1930s

Dirlik, Arif, *Anarchism in the Chinese Revolution* (Berkeley: University of California Press, 1991).

Jui, Li, *The Early Revolutionary Activities of Comrade Mao Tse-tung* (White Plains, N.Y.: M. E. Sharpe, 1977).

Saich, Tony, and Hans J. van de Ven, eds., *New Perspectives on the Chinese Communist Revolution* (Armonk, N.Y.: M. E. Sharpe, 1995).

Stranahan, Patricia, *Underground: The Shanghai Communist Party and the Politics of Survival, 1927–1937* (Lanham, Md.: Rowman & Littlefield, 1998).

van de Ven, Hans J., *From Friend to Comrade: The Chinese Communist Party, 1920–1927* (Berkeley: University of California Press, 1991).
Yang, Benjamin, *From Revolution to Politics: Chinese Communists on the Long March* (Boulder, Colo.: Westview Press, 1990).

1940s

Apter, David E., and Tony Saich, *Revolutionary Discourse in Mao's Republic* (Cambridge: Harvard University Press, 1994).
Dai Qing, *Wang Shiwei and "Wild Lilies": Rectification and Purges in the Chinese Communist Party, 1942–1944* (Armonk, N.Y.: M. E. Sharpe, 1994).
Goldman, Merle, *Literary Dissent in Communist China* (Cambridge: Harvard University Press, 1967).
Levine, Steven I., *Anvil of Victory: The Communist Revolution in Manchuria, 1945–1948* (New York: Columbia University Press, 1987).
Pepper, Suzanne, *Civil War in China: The Political Struggle, 1945–49* (Berkeley: University of California Press, 1978).
Selden, Mark, *China in Revolution: The Yenan Way Revisited* (Armonk, N.Y.: M. E. Sharpe, 1995).
Watson, Andrew, *Mao Zedong and the Political Economy of the Border Region* (Cambridge: Cambridge University Press, 1980).
Wylie, Raymond F., *The Emergence of Maoism: Mao Tse-tung, Ch'en Po-ta, and the Search for Chinese Theory, 1935–1945* (Stanford: Stanford University Press, 1980).

People's Republic of China

Barmé, Geremie R., *Shades of Mao: The Posthumous Cult of the Great Leader* (Armonk, N.Y.: M. E. Sharpe, 1996).
Chan, Anita, Richard Madsen, and Jonathan Unger, *Chen Village under Mao and Deng*, updated ed. (Berkeley: University of California Press, 1992).
Cheek, Timothy, and Tony Saich, eds., *New Perspectives on State Socialism in China* (Armonk, N.Y.: M. E. Sharpe, 1997).
Davies, Gloria, ed., *Voicing Concerns: Contemporary Chinese Critical Inquiry* (Lanham, Md.: Rowman & Littlefield, 2001).
Landsberger, Stefan, *Chinese Propaganda Posters: From Revolution to Modernization* (Armonk, N.Y.: M. E. Sharpe, 1995).
MacFarquhar, Roderick, *The Origins of the Cultural Revolution*, vols. 1–3 (New York: Columbia University Press, 1974–97).
Meisner, Maurice, *Mao's China and After* (New York: Free Press, 1986).
Schoenhals, Michael, ed., *China's Cultural Revolution, 1966–1969: Not a Dinner Party* (Armonk, N.Y.: M. E. Sharpe, 1996).

Schrift, Melissa, *Biography of a Chairman Mao Badge: The Creation and Mass Consumption of a Personality Cult* (New Brunswick, N.J.: Rutgers University Press, 2001).

Yan Jiaqi and Gao Gao, *Turbulent Decade: A History of the Cultural Revolution*, trans. and ed. D. W. Y. Kwok (Honolulu: University of Hawaii Press, 1996).

Acknowledgments

Geremie R. Barmé. "Shades of Mao." From *Shades of Mao: The Posthumous Cult of the Great Leader.* Copyright © 1996 by Geremie R. Barmé. Reprinted by permission of M. E. Sharpe, Inc., Armonk, NY 10504.

Jack Belden. "Land and Revolution." From *China Shakes the World* by Jack Belden. Copyright © 1949 by Jack Belden. Reprinted by permission.

Michael Bullock and Jerome Ch'en. "Snow." From *Mao and the Chinese Revolution,* translated and edited by Michael Bullock and Jerome Ch'en. Oxford University Press, London, 1965. Reprinted by permission.

Central Committee on the Chinese Communist Party. "Some Questions on Party History, June 1981." Translation from *Beijing Review,* July 6, 1981. Foreign Language Bureau, Beijing Review Publishing Company.

Boyd Compton. "Resolution of the Central Committee of the Chinese Communist Party on Methods of Leadership, June 1, 1943." From *Mao's China: Party Reform Documents 1942–1944,* edited and translated by Boyd Compton. Copyright © 1952 by Boyd Compton. Reprinted by permission of the University of Washington Press.

Nick Knight. "Mao Zedong's 'Sinification of Marxism.'" From *Marxism in Asia,* edited by Colin Mackerras and Nick Knight. Copyright © 1985 by Colin Mackerras and Nick Knight. Reprinted with the permission of St. Martin's Press.

Li Zhisui. "The Emperor of Zhongnanhai, 1994." From *The Private Life of Chairman Mao,* translated and edited by Tai Hung-chao. Copyright © 1994 by Tai Hung-chao. Reprinted with permission of Random House, Inc.

Roderick MacFarquhar, Timothy Cheek, and Eugene Wu. "Talks at the Beidaihe Conference, August 1958." Excerpts from *The Secret Speeches of Chairman Mao: From the Hundred Flowers to the Great Leap Forward,* translated and edited by Roderick MacFarquhar, Timothy Cheek, and Eugene Wu. Copyright © 1989 by the Harvard University Asia Center. Reprinted by permission.

Mao Zedong. "On the Correct Handling of Contradictions among the People, June 1957." Translated from *Selected Works of Mao Tse-tung.* Foreign Language Press, 1977. "Bombarded the Headquarters." (New China News Agency, August 4, 1967). Translated from *Quotations from Chairman Mao Tse-tung,* Foreign Language Press, 1968.

Bonnie S. McDougall. "Talks at the Yan'an Conference on Literature and Art: A Translation of the 1943 Text with Commentary." Copyright © 1980 by the Center for Chinese Studies, The University of Michigan. Reprinted with permission.

Rae Yang. "At the Center of the Storm." From *Spider Eaters: A Memoir.* Copyright © 1997 The Regents of the University of California. Reprinted by permission of the University of California Press.

Michael Schoenhals. Excerpt from *China's Cultural Revolution 1966–1969: Not a Dinner Party.* Reprinted by permission of M. E. Sharpe, Inc., Armonk, NY 10504.

Stuart R. Schram and Nancy J. Hodes. "Report on the Peasant Movement in Hunan, February 1927." From *Mao's Road to Power: Revolutionary Writings, 1912–1949: Volume II: National Revolution and Social Revolution, December 1920–June 1927,* translated and edited by Stuart R. Schram and Nancy J. Hodes. Copyright © 1994. Reprinted by permission from M. E. Sharpe, Inc., Armonk, NY 10504.

Stuart R. Schram and Nancy J. Hodes. "On New Democracy, January 15, 1940." From *Mao's Road to Power: Revolutionary Writings, 1912–1949: Volume VII, 1940–1941,* translated and edited by Stuart R. Schram and Nancy J. Hodes. Reprinted by permission from M. E. Sharpe, Inc., Armonk, NY 10504.

Stuart R. Schram. "The Chinese People Have Stood Up, September 1949," "American Imperialism Is Closely Surrounded by the Peoples of the World, 1964." From *The Political Thought of Mao Tse-tung,* Second Edition, translated and edited by Stuart R. Schram. Copyright © 1969 Praeger Publishers, a division of Greenwood Publishing Group. Reprinted by permission.

Stuart R. Schram. "Just a Few Words (Talk at the Central Work Conference), October 25, 1966," "The Struggle on Two Fronts, 1967." From *Mao Tse-tung Unrehearsed: Talks and Letters 1956–1971,* translated by Stuart R. Schram. © 1967 Penguin Books Ltd. Reprinted by permission.

Edgar Snow. "Interview with Mao." From *Red Star Over China* by Edgar Snow. Copyright © 1968 by Edgar Snow. Used by permission of Grove/Atlantic, Inc.

Jeffrey Wasserstrom. "Mao Matters." From *China Review International,* vol. 3, no. 1 (Spring 1996). Copyright © 1996 University of Hawaii Press. Reprinted with permission.

Index

Note: Chinese names appear with family name first: Mao Zedong is "Mr. Mao" and is indexed under "M."

absolute power, 25, 30, 221, 223
accumulation, by cooperatives, 144–45
Acton, Lord, 25
Africa, 159
African Americans, 187
Agrarian Revolution, 82, 107–8, 218
agricultural colleges, 161
agricultural cooperatives, 143–45, 234
  accumulation by, 144–45
  disturbances by, 156
  "Little Red Book" quotations on, 176
agricultural production
  backyard steel furnaces and, 208
  deep ploughing and, 160, 161–62
  five year plans and, 161
  inflated figures for, 25, 143, 208–9
  laborers, 98
  vegetable farmers, 66–67
airports, 160, 164
Analects (Confucius), 189
anarchy, 133
animal food, 69
Anshan Steel Mill, 22*f*, 163
antagonistic contradictions, 130. *See also* contradictions
  defined, 130
  resolution of, 136
anti-Americanism, 167
anti-communism, 107–8
anti-feudalism, 104, 105, 110–12. *See also* feudalism
anti-imperialism, 104, 105, 110–12. *See also* imperialism
Anti-Japanese Theater, 185–86
Anti-Japanese United Front, 90, 92, 94, 104
Anti-Rightist Campaign, 24, 25, 129, 234
Apter, David, 33, 219–25
aristocracy. *See* gentry

armed forces. *See also* Eighth Route Army; New Fourth Army; People's Liberation Army; Red Army
  irregular household militia, 60
  of landlords, overthrow of, 59–60
  of peasants, establishment of, 59–60
  standing household militia, 60
art
  freedom of thought in, 150
  revolutionary work and, 113–17
  social role of, 113
  subordination to politics, 113, 116–17
Asia, 159, 168
atomic bomb, 234
"At the Center of the Storm" (Rae Yang), 210–15
Austria-Hungary, 105
authorities
  clan, 62–66, 175
  masculine (husband), 11, 62, 64, 175
  overthrow of, by peasant associations, 62–66
  of peasant associations, 43–44
  political, 62, 175
  religious, 175
  types of, 175
Autumn Harvest uprising, 5

backyard steel furnaces, 205–6, 208–9
baihua, 9, 103*n*. *See also* Chinese language
bailiffs, overthrow of, 60–62
banditry
  elimination of, 71–72
  peasant movement and, 53
bans, by peasant associations, 67–71
Bao'an, 185
Barmé, Geremie
  Shades of Mao, 226–31
beef, 69–70
beggar-bullies, 70

245

historical background, 76
Hundred Flowers Campaign and, 23–24, 128–29, 149–55, 233, 234
"Little Red Book" quotations on, 172–73
long-term coexistence and mutual supervision among political parties, 154–55
Mao's alienation from, 22–23, 26–27
Mao's influence on, 33, 216–19
Mao's leadership of, vi, 1
Mao's writings and, 39–40, 216–19
May Fourth Movement and, 105
newspaper support of, 107
party cell organization, 10
party unity, 135
persecution of, 108
political opposition and, 33–34
reactionary campaigns against, 107–8
resolution on leadership methods, 112–17
resolution on the role of Mao, 33
support for, 103–4
Three People's Principles and, 99
Chinese Culture (journal), 78
Chinese language
biahua, 9, 103*n*
changes in, 103
romanization, vii–viii
vernacular, 9, 103*n*
writing system, 9
"Chinese People Have Stood Up, The" (Mao Zedong), 125–27
Chinese Political Consultative Congress, 125
Chinese Sciences, 32
Chongqing, 5
civil rights, 133
civil service examinations, 9, 232
clan authority
"Little Red Book" quotations on, 175
overthrow of, by peasant associations, 62–66
class struggle, 31–32
during Chinese civil war, 17–18
contradictions and, 140–41
development of Marxism through, 150–52
Hunan peasant uprisings and, 41
"Little Red Book" quotations on, 174, 177
social contradictions and, 131
universal laws and, 199
coercion, avoiding, 134, 136
coexistence
with imperialist countries, 159
of parties within China, 154–55
collectivization of agriculture, 143–45, 156, 176, 234
colonialism, 80, 88, 90
Comintern (Communist International organization), 10
commercialism, 34*f*
communal dining, 163

communes. *See* people's communes
communism
labor and, 163
as long-term goal, 77
opposition to, 96–97
people's communes and, 163
restricting, 94
in Soviet Union, 164–65
Communist Party. *See also* Chinese Communist Party (CCP)
legality of, in capitalist countries, 133–34
comprador, 151
Confucianism, 8, 100, 227*n*
New Culture Movement and, 9–10
Confucius, 47*n*
Congo, 167–68
congress system, 91
consumers' cooperatives, 74
contradictions
antagonistic (with enemy countries), 130–41
co-operative transformation of agriculture, 143–45
correct handling of, 129–59
counter-revolutionaries, 141–43
democratic resolution of, 134–35
denial of, 138
distinguishing social and antagonistic, 137
emergence of, 128
Hundred Flowers Campaign, 149–55
immunity and, 129
industrial, 145–46, 158–59
intellectuals, 146–48
lessons from, 157–58
"Little Red Book" quotations on, 174
minority nationalities, 148–49
Red Guards and, 211
resolution of, 128–29, 131, 138
in socialist *vs.* capitalist countries, 138
social (within China), 130–41
strikes and disturbances, 155–57
types of, 130–41
unity of opposites law and, 137–38
cooperatives, 74
accumulation by, 144–45
agricultural, 143–45, 156, 176, 234
consolidation of, 144
credit, 71, 74
industrial, 163–64
corruption
in county government, peasant associations and, 60–62
peasant attacks against, 43–45
power and, 25, 30
counter-revolution
elimination of counter-revolutionaries, 141–43
fear of, 229
county magistrates, overthrow of, 60–62
Coup of 1898, 81
Cow Demons and Snake Spirits, 229

Printed in the United States
By Bookmasters